W9-AXF-450

REFLECTIONS ON BIG SCIENCE

REFLECTIONS ON BIG SCIENCE

Alvin M. Weinberg

THE M.I.T. PRESS

Massachusetts Institute of Technology
Cambridge, Massachusetts, and London, England

Preface

During the eighteen years in which I have been Research Director and Director of the Oak Ridge National Laboratory, I have made more than my share of speeches. Some of the speeches have been, basically, justifications for the large sums of public money that pour into modern science. Some have been justifications for the institutions of Big Science, particularly the national laboratories — justifications which probably will have to be stated and restated until the laboratories, now barely twenty years old, have acquired the sanction of antiquity enjoyed by the universities.

Concern with these topics is, I believe, a proper business of a laboratory director. After all, as Dr. E. R. Piore of IBM once told me, the first job of a laboratory director is to assure continuing and ample support for the institution he directs. And in about half the talks in this volume (those collected under the titles "The Promise of Scientific Technology" and "The Institutions of Big Science"), I argue that the large mission-oriented laboratories and what they do are indispensable to the society that supports them.

The other talks, grouped under the titles "The Choices of Big Science" and "The Problems of Big Science," have a rather different origin. I have served on the Scientific Advisory Board to the Air Force, the President's Science Advisory Committee, and the Committee on Science and Public Policy of the National Academy of Sciences. Now committees, in my view, can no more produce wisdom than they can design a camel. The atmosphere of a committee is too competitive, too verbal, too formal for wisdom to sprout there. Wisdom is a very personal kind of thing; it flourishes best when a single mind thinks quietly and consistently — more quietly and consistently than is possible when one is engaged in the rough-and-tumble of committeeship with its often tendentious and personal exchanges. Thus, I have felt that some of the most troublesome questions underlying the work of the President's Science Advisory Committee — such as the allocation of resources among competing scientific fields, or the allocation of resources between science as a whole and other public enterprises, or, perhaps the most difficult of all, whether the new style of Big Science is blunting science as an instrument for uncovering new knowledge — ought to be thought through by individuals who would then set their thoughts down in essays. Out of many such essays, written by different people, could come, if not clarity and guidance, at least a common language and framework in which to conduct the discourse. The Committee could then serve to criticize, or even to satisfy egos, rather than to create more wisdom than was contained in the individual essays. I am very pleased that George B. Kistiakowsky, former Chairman of the National Academy of Sciences Committee on Science and Public Policy, used this technique of combining personal essays into a report, "Basic Research and National Goals," which appeared in March of 1965.[1]

I have no illusions about how much influence any one set of observations can or should have on such complex issues. I hope that my essays contribute to clarifying the issues and, perhaps more important, that they help catalyze the debate both among those who agree and among those who disagree

[1]"Basic Research and National Goals," A Report to the Committee on Science and Astronautics, U.S. House of Representatives, by the National Academy of Sciences, U.S. Government Printing Office, Washington, D.C. (March 1965).

with me on the relation between modern society and modern science.

Most of the task of reshaping these talks so as to make them into a loosely coherent collection of essays was performed at the University of California at Santa Barbara. I am grateful to that institution for allowing me, during a stay as a Regents' Lecturer, to devote my time to preparing these talks for publication. I am also much indebted to my administrative assistant, Mrs. Elizabeth B. Richardson, who has helped so selflessly in the preparation of the manuscript; to my secretary, Mrs. Peggy Lyons; and to my many colleagues at the Oak Ridge National Laboratory, whose collective wisdom has largely been distilled in these talks. Finally, I wish to thank my wife, who has been a most constructive critic and understanding informal collaborator, both when I first presented these talks and when, at Santa Barbara, I reworked them for publication.

Oak Ridge, Tennessee   ALVIN M. WEINBERG
August 18, 1966

Acknowledgment

I wish to acknowledge the cooperation of the editors of the following publications in granting permission for republication of various articles which first appeared in their journals: *Bulletin of the Atomic Scientists, Chemical and Engineering News, Minerva, Physics Today, Science, Science and Technology,* and the *SSRS Newsletter.*

ALVIN M. WEINBERG

Contents

I   THE PROMISE OF SCIENTIFIC
    TECHNOLOGY: THE NEW
    REVOLUTIONS   1

II THE PROBLEMS OF BIG SCIENCE:
    SCIENTIFIC COMMUNICATION   39

III THE CHOICES OF BIG SCIENCE   65

    1. Criteria for Scientific Choice   65
    2. Criteria for Scientific Choice II: The Two Cultures   85
    3. The Coming Age of Biomedical Science   101
    4. Scientific Choice and Human Values   115

IV THE INSTITUTIONS OF BIG SCIENCE   123

    1. National Laboratories and Missions   126
    2. Universities and Disciplines   145
    3. The Federal Grant University and the Federal
       Laboratory   164

    INDEX   175

I

THE PROMISE OF
SCIENTIFIC TECHNOLOGY:
THE NEW REVOLUTIONS[1]

Our federal government spends $16 billion per year
on research and development. At least half of this is spent
for military development; about $5 billion for space explora-
tion, $1 billion for health research, $500 million for atomic
energy research, and the rest for many smaller scientific and
technical enterprises.

The institution to which I belong, the Oak Ridge National
Laboratory, spends about $100 million per year. Originally
we were concerned entirely with atomic energy, though in
recent years we have expanded our scope to include such
matters as desalting the sea and civil defense. Many of us
whose scientific lives have been involved with one or another
of these huge scientific enterprises often ask ourselves, "Is
society getting its money's worth from what we, and other
Big Scientists, spend?" It will be my purpose in this essay
to explain why society could hardly survive for many more
generations without the fantastic developments that have

[1]This essay contains material that I first presented in "Energy as an
Ultimate Raw Material," *Physics Today 12*, 18–25 (November 1959);
in "Today's Revolution," *Bulletin of the Atomic Scientists XII*, 299–
302 (October 1956); and in "Effects of Scale on Modern Science and
Technology," *Society for Social Responsibility in Science Newsletter*
(November 1963).

come out of Big Science. The whole future of our society depends upon the continued success of our science and our scientific technology.

In pursuing this theme, I shall call mostly upon my experience in atomic energy. I do this for two reasons: first, because most of what I know in science and technology relates to atomic energy, and second, because voices have been raised — for example, David E. Lilienthal's in his *Change, Hope, and the Bomb*[2] — casting doubt on the validity of the whole nuclear enterprise. I think it is important that those of us who see in nuclear energy, and in the other marvels of modern science, a means to achieve H. G. Wells's world set free[3] ought to speak out. Our vision of an abundant world is well worth striving for, and it would be even more vigorously sought if only society at large had a clearer idea of the shape of that world.

The Thermodynamic Revolutions

The overriding concerns of society in the next few generations must be the question of peace and the question of population. I shall set aside the first question for the moment and consider only the second. The next twenty-five years may see the world's population rise from its present 3.2 billion to 6 billion. We "should do well to ponder the significance of this development in terms of the destiny of our species.

"These next twenty-five years form part of a process which began some 200,000 years ago and which is about to culminate in man's full possession of the earth."[4]

The most obvious threat of uncontrolled population growth

[2]Princeton University Press, Princeton, New Jersey (1963).

[3]*The World Set Free: A Story of Mankind,* Dutton and Company, New York (1914).

[4]*Population Bulletin XV,* 21, Population Reference Bureau, Inc., Washington, D.C. (March 1959).

2

is the threat foreseen by Thomas Malthus: population grows faster than do the means of subsistence. Science, at least in the West, has thus far forestalled the consequences of Malthus' dilemma. It has created abundance, and in the United States, the problem-laden affluent society. The question is whether science can continue to maintain the living standard until we learn how to control the population explosion. *Everything I say is therefore predicated on the assumption that we shall eventually control population, hopefully at a size not much more than twice the current population. If we cannot control the growth of population, nothing can save us.*

Malthus' dilemma is, from one point of view, an imbalance between the energy available to man and the energy he requires. For, as I shall explain later, with energy we can convert common materials into the necessities of life: we can convert sea water into fresh water, or nitrogen in the atmosphere into nitrate fertilizer (and ultimately into food), or even coal into gasoline (by hydrogenation with hydrogen obtained from electrolysis of water).

Malthus overlooked a second dilemma. This has to do with the increase in complexity, in the proliferation of the semantic environment, which accompanies the growth of population. As the number of people in a given location grows, the number of semantic contacts between people also grows. In simplest approximation, the number of contacts grows as the square of the number of people. Life becomes more complicated. There are more people to generate ideas, social contacts, personal interactions. The technology of mass dissemination imposes these stimuli upon us with alarming effectiveness. Our newspapers, not to say our scientific journals, get thicker. Our media of communication, including our transportation system, are stretched ever harder. Each individual is exposed to many more sensory impressions than was his father or his grandfather. But our ability to absorb sensory impressions hardly grows: each person merely can know less of what goes on around him, can interact less efficiently with the rest of society. We become specialized in outlook. We experience the same frustration that is felt by the older scientist who, once knowing the whole of a scientific field, must now content himself with knowing a tiny part of it.

A striking example of this "proliferation of complexity" is the telephone system. Whenever a new subscriber is added, the telephone company must add another phone to the system. At the same time, the company must expand the central switching system by a much larger increment so that the new subscriber can communicate with every other subscriber. The complexity of the switching system expands (just as do contacts between individuals) much faster than does the system itself; finally, the switching exchange dominates the whole system. This impasse was anticipated by the telephone companies as early as the 1900's, and led to the introduction of automatic dialing systems. Were it not for automatic dialing, our entire population would eventually consist of telephone operators.

That something like Malthus' second dilemma operates in isolated, overcrowded animal communities is suggested by experiments[5] on the crowding of rats. When well-fed rats are crowded beyond a certain point, they tend to become withdrawn from each other. This manifests itself, among other ways, in a marked reduction in sexual activity, and a consequent reduction of the birth rate.

This second Malthusian dilemma, the dilemma of complexity, is an imbalance between the rate at which semantic stimuli — that is, information — are generated, and the rate at which the individual can respond to the stimuli. It is an *information* crisis, in contrast to the *energy* crisis that characterizes the first Malthusian dilemma.

Ordinarily we think of energy and information as being

[5]V. C. Wynne-Edwards, "Self-Regulating Systems in Population of Animals," *Science 147*, 1543–1548 (1965).

unrelated. Yet they are subsumed in the same scientific discipline, the science of thermodynamics. In order to make this connection clearer, I shall have to give a one-paragraph digression on classical thermodynamics.

Energy is the subject of the first law of thermodynamics. This law says that energy can be neither created nor destroyed, only transformed from one form into another. Entropy is the subject matter of the second law of thermodynamics. The second law says that the entropy of the universe always increases. In more· ordinary language, and somewhat inexactly, this is equivalent to saying that the disorder of the universe always increases. We know that things left to themselves decay and disorganize: a house will become unkempt unless tidied every day; weeds will ruin a garden unless they are dug out regularly; heat will flow toward cold unless work is done to reverse the flow. Thermodynamics describes this natural trend toward disorder by saying that the entropy of each of these systems has increased.

The connection between all this and the notion of information was first shown by L. Szilard in 1929.[6] He demonstrated that the information content of a system was the negative of the system's entropy; thus the second law of thermodynamics can be paraphrased: the information content of the universe decreases. In this sense, information can be viewed as the subject matter of the second law of thermodynamics; and Malthus' second dilemma, insofar as it is an information imbalance, is concerned with the second law of thermodynamics, just as Malthus' first dilemma, being an energy imbalance, is concerned with the first law of thermodynamics.

To my mind this statement of Malthus' dilemmas in the language of thermodynamics provides a neatly unified view of the human condition: the future of mankind is destined

[6]L. Szilard, "Über die Entropieverminderung in einem thermodynamischen System bei Eingriffen intelligenter Wesen," *Zeitschrift für Physik 53*, 840–856 (1929).

to be a struggle between increasing population on the one hand, and dwindling resources of energy and inability to cope with complexity on the other. But this neat view would be without substance were it not that the two major scientific and technical revolutions of our time are also concerned with energy and information. The energy revolution has suddenly presented us with completely new resources of energy, and should therefore help us to escape from Malthus' first dilemma; at the same time, it will present us with new problems. The information revolution has presented us with new ways of dealing with complexity, and should therefore help us to escape from Malthus' second dilemma, but it also will present us with new problems. Much of this essay will be devoted to examining how these two major scientific-technical revolutions, particularly the energy revolution, can be expected to ameliorate the human condition and how, in some respects, these revolutions will aggravate it.

Energy: The Ultimate Raw Material

I now consider energy and its impact on man. Many books have been written on the subject, most notably by Palmer Putnam and by Hans Thirring.[7] Writers on energy divide into optimists and pessimists. The pessimists hold that cheap and abundant energy is important but not terribly important. After all, only 2 per cent of the Gross National Product of the United States went into generation of electricity in 1964; even if the cost of electricity went to zero, the effect on the entire economy, from this viewpoint, would be minimal. The optimists, such as Harrison Brown,[8] whose views have

[7]Palmer Putnam, *Energy in the Future,* D. Van Nostrand Company, Inc., New York (1953); Hans Thirring, *Energy for Man,* Indiana University Press, Bloomington, Indiana (1958). See also Alvin M. Weinberg, "Energy as an Ultimate Raw Material," *Physics Today 12,* 18–25 (November 1959).

[8]Harrison Brown, James Bonner, and John Weir, *The Next Hundred Years,* The Viking Press, New York (1957).

strongly influenced my own, look upon cheap and abundant energy as central: as the means of resolving Malthus' first dilemma. They view a world with inexhaustible cheap energy as did H. G. Wells — as a "world set free."

I shall expound the optimists' position. First I point out, as did Harrison Brown, that as we exhaust our richer natural resources, we shall have to use more and more energy to extract the necessities of life from common materials: from rock, from the sea, from the air. Thus, although energy now accounts for but 2 per cent of our Gross National Product, it seems likely that this fraction will increase greatly in the future. Only if we can maintain, and indeed increase, our supply of very cheap energy can we hope to stave off the consequences of Malthus' first dilemma in the face of our increasing population. This point was stressed by Sir Charles Darwin in his book *The Next Million Years*.[9] Sir Charles took a very dim view indeed of mankind's future unless we discovered a very cheap and inexhaustible source of energy. I shall therefore describe how the world could extract means of subsistence from ordinary materials reasonably economically, but only if it had cheap and inexhaustible energy. Second, I shall describe the astonishing progress that has taken place, some in the past couple of years, in achieving cheap and inexhaustible energy. My major contention is that although the energy revolution envisaged by H. G. Wells when he wrote *The World Set Free* has all but arrived, we have not yet responded fully to this revolution.

The Importance of Cheap Electricity

Suppose that we have learned to produce electricity anywhere in the world at a price — say, 1.5 mills/kwh — that is as low as the cheapest electricity now available in a few very isolated hydroelectric sites, such as Rjukan in Norway. This

[9]Doubleday, Garden City, New York (1953).

price is about half the price of electricity produced in the best modern publicly owned American coal-powered stations, with coal costing $4/ton, and about four times lower than the cost of electricity in most other parts of the world, where coal costs $10/ton. I shall first show how cheap energy can be converted into the major material requirements of life: water, food, metals, and a tolerable environment.

Consider water. To extract fresh water from the sea by distillation requires a minimum of about three kilowatt hours (kwh) of mechanical work per 1000 gallons. This minimum is achieved if the process is conducted infinitely slowly. In actual practice the process requires ten to fifty times as much work. The simplest and best established process is multiple-effect distillation. In this process sea water is boiled, and the vapor, in condensing, boils additional sea water at lower pressure. This process is repeated successively, sometimes as many as thirty times. By using the heat from the condensing vapor many times, one saves energy; however, one pays in complication of the distilling apparatus. In the United States, several demonstration plants have been distilling sea water for years. The largest such plant, at Point Loma, California, produces 1.4 million gallons of fresh water per day, enough to supply an American town of about 5000 inhabitants.

Distilling sea water is no trick; the problem is to distill it economically — say, for less than 25¢/1000 gallons if the fresh water is to be used by a municipality, or for less than 5¢/1000 gallons if it is to be used generally for irrigation. The cost of water from a desalting plant is made up of two major components: capital cost and energy cost. If the energy is expensive, it pays to save energy by using a complicated, multistage still in which energy gained by condensation in one stage boils water in the next stage. If the energy is cheap enough, it pays to waste energy by using a cheap still with very few stages. This latter possibility has emerged in the past few years, largely from the work of R. Philip Hammond, formerly at Los Alamos and now at the Oak Ridge National

8

Laboratory. Hammond originally envisaged huge nuclear electrical desalting installations producing 10^9 gallons of water per day and 4000 megawatts of electricity, although he has now designed somewhat smaller units as well. In these dual-purpose plants, heat to energize the evaporators is drawn off from the low-pressure end of the turbines. Since most of the heat so drawn off would be wasted anyhow, it can be provided to the evaporators for almost nothing. Moreover, as will be discussed later, because the installations are large, their *unit* capital costs ought to be very low, although of course the entire plant will be very expensive. Hammond estimates that such large dual-purpose plants operated publicly could produce water for around 15¢/1000 gallons and by-product electricity for about 1.5 mills/kwh.

Hammond's ideas have created a sensation among people interested in desalting and in nuclear energy. President Johnson has launched a Water-for-Peace program to exploit these possibilities and to share our knowledge with water-hungry countries throughout the world. We have already cooperated with the U.S.S.R. in an exchange of information; we are now working with Israel, Mexico, and other arid countries in further exploring nuclear desalting. I have little doubt that within the next decade we shall see several large dual-purpose electricity and desalting plants springing up in arid places bordering the sea. At first these plants will produce water only for municipal or industrial use. As experience is gained in the operation of such large plants, the unit cost ought to fall, until eventually, I believe, water cheap enough for at least some agriculture will be feasible.

I turn next to food, where a primary *technical* problem is to convert energy into fertilizer — that is, into fixed nitrogen, potassium salts, and superphosphate. R. E. Blanco and others of the Oak Ridge National Laboratory have studied this question,[10] and I shall quote some of their results. If the world

[10] R. E. Blanco, J. M. Holmes, R. Salmon, and J. W. Ullmann, "An Economic Study of the Production of Ammonia Using Electricity

were to use fertilizer at the same per capita rate as we in the United States use it, consumption of fixed nitrogen, phosphate, and potassium would increase from the 27 million metric tons used in 1961 to 181 million metric tons by the year 2000. Blanco estimates that the cost of energy needed to fix nitrogen from the air by the arc process, with electricity at 1.5 mills/kwh, is only about 4¢/pound of nitrogen. The total cost of fixed nitrogen, which includes capital and operating costs as well as the cost of energy, might then be as little as 11¢/pound. This is about 50 per cent higher than the present cost of nitrogen, as ammonia, obtained from natural gas and air, though no more expensive than nitrogen from Chile saltpeter. The total cost of the estimated 80 million tons of nitrogen needed by the year 2000, even at 11¢/pound, would be about $18 billion per year, or about $4 for the nitrogen needed to fertilize the crops necessary to feed one person per year.

Potassium salts can in principle be extracted from the sea as a by-product in a sea-water distillation plant. Such processes for extracting potassium are not very economical as yet, but Blanco is optimistic that they can be developed. As for superphosphate, electricity is already used by the Tennessee Valley Authority (TVA) to produce superphosphate from raw phosphate rocks. In 1950 Schurr and Marschak[11] calculated that if electricity were available at 3.2 mills/kwh, superphosphate could be made economically from raw Florida phosphate rocks by electrical rather than by chemical methods. If electricity were cheaper, presumably poorer grade phosphate rocks could be used economically. Raw phosphates are distributed quite widely, and although they seem to be scarce in China and India, the world is well enough endowed

from a Nuclear Desalination Reactor Complex," ORNL–3882, Oak Ridge National Laboratory, Oak Ridge, Tennessee (June 1966).

11S. H. Schurr and Jacob Marschak, *Economic Aspects of Atomic Power*, Part Two, Chapter VI, "Phosphate Fertilizers," 124–134, Princeton University Press, Princeton, New Jersey (1950).

with phosphate rock in many places to supply this raw material for a very long time.

To summarize, with electricity at 1.5 mills/kwh and using as raw material only air, sea water, and phosphate rock, we probably could produce fertilizer that is only about 50 per cent more expensive than the cheapest fertilizers now available. Moreover, this source of fertilizer is essentially inexhaustible. In this sense, cheap electricity could indirectly provide sufficient food to keep up with the population, at least for a considerable time.

Conversion of electricity into metals is a similar story. All the important metals appear in nature as oxides, the metals having lost their valence electrons. To obtain metals from ores, one must supply electrons. In the smelting of iron ore, electrons are supplied by coke. In principle, electrons can be supplied directly by electricity, or less directly by hydrogen, which is produced from the electrolysis of water. The direct cost of the energy is negligible; at 1.5 mills/kwh, the cost of energy would still add only about two tenths of a cent to the price of a pound of iron. Unfortunately, the needed technology for direct reduction of iron is not developed (although electrolytic reduction of aluminum is a well-developed art). However, electric furnaces in which electricity supplies heat to a mixture of low-grade coke and iron ore have been used to reduce iron ore on a fairly big scale. The advantage of such furnaces is that they use a low-grade, generally abundant coal rather than the high-grade coking coal needed for blast furnaces.

Eventually we shall have to get metals from lower and lower grade ores. Harrison Brown examined this matter several years ago.[12] He concluded that the cost of the energy

[12]H. Brown and L. T. Silver, "The Possibilities of Securing Long Range Supplies of Uranium, Thorium and Other Substances from Igneous Rocks," *Proceedings of the International Conference on the Peaceful Uses of Atomic Energy 8,* 129–132, United Nations, New York (1956).

needed for crushing and hauling rocks from which very low-grade metallic oxides could be extracted was high but not intolerable. He estimated that the energy from about 25 kilograms of coal would be needed to process a ton of granite, but that from this ton of granite one could eventually extract 70 kilograms of aluminum, 10 kilograms of iron, 4 grams of uranium, and 13 grams of thorium. Of course, one would not use common rock as raw material for a long, long time; however, it is reassuring that Brown's estimates suggest that we can get metals, particularly uranium and thorium, from the rocks if only our energy is cheap enough.

If one considers supplying electrons directly to reduce metallic ores to their metals and supplying electrons via elemental hydrogen, the latter is probably the more promising. If hydrogen is sufficiently cheap, it can be used to advantage not only to win metals from ores but also to fix nitrogen as ammonia, or to convert coal into liquid fuels. Thus, the full utility of cheap energy as the ultimate raw material for heavy chemical processing may depend strongly upon our devising cheap ways of electrolyzing water into free hydrogen and oxygen. Here the needed processes seem to have received an unexpected boost from both military and space technology. Compact fuel cells for spacecraft have been developed in which current densities at the electrode approach 1000 amperes per square foot. This is about five times the current density achieved in large-scale electrolytic cells now used for manufacture of hydrogen. If these high-current-density electrodes could be applied to large-scale electrolysis, presumably the unit capital costs could be drastically lowered, and the way to cheap hydrogen via cheap electricity, and then to reasonably priced metals, fertilizer, and liquid fuel, would be fairly clear.

Finally, I mention energy for space heating. Again I draw upon Brown, who estimates that eventually two thirds of the world's space heating will be supplied by solar heat, the rest being provided by electricity. In the TVA area, where elec-

TABLE 1

PROJECTED ENERGY INPUT PATTERN FOR YEAR 2060
(After Brown, Bonner, and Weir[8])
(World Population 7×10^9)

Source	Equivalent Metric Tons of Coal (Billions)	Equivalent Heat Energy	
		10^{18} Btu*	Mwy† Heat
Solar energy (for 2/3 of space heating)	15.6	.42	140×10^5
Hydroelectricity	4.2	.10	38×10^5
Wood for lumber and paper	2.7	.07	24×10^5
Wood for conversion to liquid fuels and chemicals	2.3	.06	21×10^5
Liquid fuels and petro-chemicals produced via nuclear energy	10.0	.27	90×10^5
Nuclear electricity	35.2	.96	320×10^5
	70.0	1.88	633×10^5

*Btu — British Thermal Unit, 252 calories.
†Mwy — Megawatt year.

tricity for heating costs only 7 mills/kwh, electrical heating is competitive with heat from coal. Of course for house heating, even if electricity were *generated* at 1.5 mills/kwh, its cost to the consumer would be much higher — say, 5 mills/kwh. Still, at 5 mills/kwh, electric heating would probably compete with heat from fossil fuel in much of the United States, and we therefore can look forward to the day when many of our houses will be heated with electricity from nuclear reactors.

I recapitulate by giving a projected energy budget for the world of 2060 drawn up by Brown, Bonner, and Weir in their book *The Next Hundred Years*.[13] These authors assume that

[13]Harrison Brown, James Bonner, and John Weir, *op. cit.*

the world's population will be 7 billions at that time, and that most of the types of conversion of energy into materials I have described, other than distilling of sea water, are feasible. If sea-water distillation is included, Brown's energy budget would probably increase by 10 or so per cent.

The total projected yearly consumption of energy, 6.3×10^7 megawatt years of heat, is the equivalent of 70×10^9 tons of coal per year, or 10 tons per person. This is about eighteen times the present equivalent energy input of 0.35×10^7 Mwy of heat. At Brown's ultimate rate, the present fossil fuel reserves of perhaps 2400×10^9 tons would hardly last fifty or so years. Thus, we must take for granted the world's ultimate dependence on some source of energy other than fossil fuel.

The Nuclear Energy Revolution: Cheap Energy

Most of the public awareness about energy has been focused on nuclear energy, and of course I shall return to its role. However, our interest in nuclear energy ought not to cause us to overlook the remarkable advances that have occurred in the technology of conventional power. Thermal efficiencies have crept up steadily each year until now new large stations operate routinely at better than 40 per cent efficiency. Units have become larger and larger, and several plants are now under construction in the United States in which a single turbine and boiler produce 1000 megawatts of electricity (Mwe). As the plants become larger, their unit capital cost falls: the new 615 Mwe Cardinal Plant of the American Electric Power Company cost $107 per kilowatt of electricity (kwe) in 1964, though a duplicate plant, built in 1966, is estimated to cost $125/kwe. Advances in transmitting electricity have also been spectacular. Voltages are up to 750 kilovolts, and Consolidated Edison Company of New York has considered transmitting 2 million kilowatts of electricity from hydroplants in Labrador, a distance of 1200 miles. The technology of mining and hauling coal is improving rapidly, possibly under pressure of competition from nuclear energy. The Joy Manufacturing Company has developed a prototype automatic coal mining machine that in a day can mine as much coal as can several hundred miners. The railroads have developed new bulk carriers — "unitrains" — which drastically reduce the cost of hauling large amounts of coal. The result of all these incremental gains is that modern privately owned steam plants in much of the United States now generate power at around 4.5 mills/kwh. If the plants are publicly owned, so that the annual charges are 7.5 per cent instead of the 13.5 per cent assessed on privately owned plants, the cost of generation would be reduced by around 0.9 mill/kwe.[14]

[14]Philip Sporn, former president of the American Electric Power Company, in a talk before the Southern Interstate Nuclear Board, Oak Ridge, Tennessee, January 20, 1966, gave the following estimate of the current economic status of coal-fired power plants.

THE PRESENT ECONOMIC STATUS OF COAL-FIRED POWER BASED ON EXPERIENCE WITH THE CARDINAL STEAM PLANT (AS OF 1966)

	Cardinal-Type Coal-Fired Unit		
Capability (Mwe)	615		
Unit Capital Cost $/kwe	125		
Heat Rate (Btu/kwhe)*	8650		
Capital Charges at 80% Load Factor, 13.5% Annual Charges (mills/kwhe)	2.4		
	20¢/MBtu†	25¢/MBtu	30¢/MBtu
Fuel (mills/kwhe)	1.73	2.16	2.60
Operating and Maintenance (mills/kwhe)	.30	.30	.30
Energy Cost (mills/kwhe)	4.43	4.86	5.30

*kwhe — kilowatt hours of electricity.

†MBtu — million Btu. Coal at $4.80/ton corresponds to heat energy at about 20¢/MBtu, assuming the energy content of the coal is 12,000 Btu/pound.

15

But nuclear energy has moved even more rapidly. As recently as 1963, the nuclear technology community was rather pessimistic about the prospects for nuclear energy's becoming competitive. One quip had it that "nuclear energy would become competitive within 10 years of the time the prediction was being made." Most of us could not see how the capital costs of nuclear plants could be reduced much below $200/kwe, and at this price, nuclear energy would not be competitive. However, since 1963 several spectacular happenings have completely changed the outlook.

The first was the announcement early in 1964 by the Jersey Central Power and Light Company that it had contracted with the General Electric Company for a 515 megawatt boiling water reactor to be built at Oyster Creek for the extraordinarily low (by nuclear standards) price of $132/kwe. If the reactor could be operated at its "stretch" rating of 620 Mwe, the unit capital cost would fall to $109/kwe, which is lower than the price of most conventional electric plants of this size. This announcement created a sensation. The sensation was compounded when, in fairly quick succession, firm price bids for many other large nuclear plants were announced: Altogether, as of the middle of 1966, our country's utilities have contracted for 21 million kilowatts of nuclear generating plants. The largest of these is the 2.2 million kilowatt TVA plant at Browns Ferry, Alabama, that is expected to generate electricity at about 2.4 mills/kwhe under TVA financing conventions. A coal-fired plant of the same size was estimated by TVA to generate electricity at 2.8 mills/kwhe. This is particularly significant since TVA lies in the heart of the Appalachian coal country and enjoys the benefit of very low-priced coal.

What happened so suddenly to make nuclear energy competitive? Of course, one must remember that all of these new plants are still to be completed. And there is a chance that the plants will not operate as well as expected. But the good operating experience of the Yankee Atomic Electric plant, a pressurized water reactor operating since 1960 at 185 Mw, and the Dresden No. 1, boiling water plant operating since 1959 at 210 Mw, puts the likelihood of failure very low. Certainly the private utility companies that have bought these new reactors are willing to invest their own money in pressurized and boiling water nuclear reactors.

Of the factors that seem to be involved in this drastic reduction of the cost of nuclear power plants, three stand out.

First, the Oyster Creek boiling water reactor is the sixth or seventh of a series of reactors of this type. It is inevitable that designers of a series of reactors, all of the same general type, find ways of improving each successive reactor. For example, the steam that is generated directly in the core of the new Dresden No. 2 boiling water reactor is separated from entrained water in the pressure vessel itself, whereas in Dresden No. 1, the steam is de-entrained in a separate, and expensive, steam drum.

Second, there is the working of the competitive market place. The market for large civil water reactors in the United States is now dominated by the General Electric Company, which favors the direct boiling water system, and by the Westinghouse Electric Corporation, which favors the pressurized water type. Both companies have bid on every major nuclear power installation. One can hardly doubt that the spirited competition between these two giants has lowered the price of the current crop of reactors.

Finally, and perhaps most important, the new reactors are all very large (the two new TVA reactors are designed to generate 1100 megawatts apiece). As has been stressed most strongly by Hammond, large nuclear reactors are much cheaper, per unit of output, than are small ones. This comes about because a nuclear reactor is, in principle, an unlimited energy source; the amount of heat that can be drawn from the reactor is limited in principle only by the size of the heat-exchange equipment and the maximum temperature of the reactor. The cost of a nuclear reactor increases as its power output increases, but not as fast as its output. The cost of the reactor itself (its instrumentation, its shield, its control room, and so on) hardly increases at all as the output of the reactor increases. The cost of the heat-exchange equipment increases with heat output, but, like most large-scale equipment, at a slower rate than the heat output itself. Thus the cost per kilowatt will fall as the output of the reactor increases. That the unit cost of nuclear reactors, like the cost of conventional power plants, decreases with increasing size is now attested to by price lists established both by General Electric and Westinghouse.

The other major component of cost in the nuclear reactor, the fuel cycle cost, also seems to fall as the size of the system increases. The fuel cycle cost is made up of four components: carrying charges for the fuel inventory, fabrication of the fuel elements, burnup of the fissile material, and chemical processing to recover unburned fuel. Most of these costs fall sharply as the scale of the operation is increased, at least for reactors that use natural uranium or reactors that efficiently convert the abundant U^{238} into Pu^{239}. Moreover, fabrication and chemical processing, not to speak of separation of the fissile U^{235} from nonfissile U^{238}, are operations that lend themselves well to mass production. If these enterprises are conducted on a large enough scale, then the costs approach more and more nearly the cost of the raw materials, chemicals, and power. Such reduction in cost has been strikingly demonstrated at the great diffusion plants in Oak Ridge, Paducah, and Portsmouth that separate U^{235} from natural uranium. Because U^{235} is separated on such an enormous scale, a gram of separated U^{235} now costs only about four times as much as its initial cost as unseparated isotope. Estimates based on demonstrated performance of the huge fabrication and reprocessing plants at Savannah River and at Hanford suggest that, if a standardized fuel element were used, and reprocessing could be done for a group of reactors producing altogether 25,000 or, better, 50,000 megawatts of heat, the fuel cycle costs for certain nuclear reactors could be as low as one tenth the fuel cost of coal. Herein lies the great economic advantage of nuclear reactor power plants as compared with fossil fuel plants. The basic fuel cost in a nuclear system is potentially extremely low; in some breeder reactors the fuel cycle cost is estimated to come to around 0.2 mill/kwhe, whereas very few coal-burning plants have fuel costs as low as 1.7 mills/kwhe.

Where do these projections finally lead? The over-all cost of energy from the Oyster Creek plant was estimated by Jersey Central to be less than 4 mills/kwhe. This estimate is based on rather low fixed costs (10.4 per cent instead of the usual 13.5 per cent), high load factor (88 per cent instead of the more usual 80 per cent), and a fuel cycle cost of 1.5 mills/kwhe, which Oyster Creek is not expected to reach until its third fuel loading has been burned. If the annual charges are taken as 13.5 per cent and the load factor is 80 per cent, the cost of energy would be around 4.4 mills/kwhe, which still is a little lower than Sporn's 1966 estimate of the cost of energy from the coal-fired Cardinal plant.

But I believe these estimates are only a beginning, and that important reductions are in sight. The largest saving should come in the fuel cycle; here, for the reasons I have already mentioned, the fuel cycle, operating, and maintenance costs ought to fall to 0.5 mill/kwhe or less. The capital cost, if the plants are even larger than the ones now being built (say 3000 Mwe), could in my opinion plausibly fall to $90/kwe. The total cost of electricity, from a privately owned plant operating at 80 per cent load factor, might then be as little as 2.5 mills/kwhe.

Nor is this the plausible lower limit. For if these very large plants were base-loaded, and particularly if they were used to supply energy for chemical processes, the load factor might be 95 per cent, not 80 per cent. Moreover, there is a good chance that plants of this sort might last much longer than the thirty years on which their amortization rate is calculated.

Once the plant is written off, the fixed charges might fall below 10 per cent, and the out-of-pocket cost for operating such plants would be very low indeed. In any case, if the plant is publicly owned, so that it pays no taxes, the annual charges would be around 7.5 per cent, and the over-all power cost would approach 1.5 mills/kwhe.

A word should be said here about the difference between the cost of electricity at the generating plant and its price to the consumer. It is true that the price to the consumer, including transmission, maintenance of distribution system, and so forth is usually much higher than the actual cost of generation. However, as the customer's load increases, the difference between generating cost and sales price diminishes. For example, the large gaseous diffusion plants in Oak Ridge, which have used as much as 2 thousand megawatts of electricity, paid essentially the generating cost to TVA for this huge block of power. Since the uses, such as large-scale chemical processing, would involve very large blocks of power, and the chemical plants would be close to the source of power so that transmission would be very cheap, 1.5 mills/kwhe is not an implausible assumption for the ultimate cost of power actually used.

When first put forward, the prediction that nuclear energy from large publicly financed plants would cost around 1.5 mills/kwhe was so astonishing and, if correct, could have such profound effect, especially on the possibility of economically desalting the sea, that the matter was reviewed by a committee in the United States representing the Atomic Energy Commission, the Department of the Interior, and the Federal Power Commission. The triagency study[15] concluded that energy from extremely large publicly owned nuclear power plants might well sell for as little as 1.6 mills/kwhe. This is the lowest price for energy from a thermal plant ever

[15]"An Assessment of Large Nuclear Powered Sea Water Distillation Plants," prepared for the Office of Science and Technology, U.S. Government Printing Office (March 1964).

projected in a responsible government study. To many in the nuclear energy community the estimate seems amply justified by the technical situation.

The Nuclear Energy Revolution: Inexhaustible Energy

How long will our sources of energy last? At the rate of consumption projected by Brown, coal, if used as a major energy source, would last only fifty or so years. The situation is even less favorable if we are to depend on the U^{235} contained in very cheap uranium ores. The entire energy content of U^{235} derived from cheap uranium ore is probably only a few per cent of the energy content of the world's coal.

However, there are two other sources of energy — one remote, the other very real — which are inexhaustible. The first is the controlled thermonuclear fusion of deuterium, or "burning the sea." The energy content of all the sea's deuterium is infinite for all practical purposes, about 10^{10} times Brown's yearly energy budget. However, in spite of much experimentation throughout the world, no one has been able to create the conditions necessary to burn deuterium in a controlled way. These conditions are formidable: a temperature of one billion degrees (at which temperature matter is converted into plasma — that is, a collection of independently moving positive ions and electrons), a pressure of 50 atmospheres held solely by a magnetic field, and a residence time of the swiftly moving deuterium ions of a second or so. Under these extreme conditions the plasma tends to be unstable; it moves wildly toward the confining walls and dissipates itself. The outlook for eventually learning how to stabilize the plasma and ultimately how to burn the sea fluctuates from year to year. At the moment physicists have learned how to eliminate the gross "macroinstabilities" of the plasmas by imposing peculiarly shaped magnetic fields, the so-called "Ioffe" fields (named after the Russian physicist who first demonstrated experimentally that such fields suppress gross

21

instabilities). However, there are a host of subtler "micro-instabilities" that must be eliminated before one can even begin to say whether controlled fusion will ever be feasible, or if it is, whether it will ever be an economical way of producing energy.

There is another, much more immediate possibility for achieving an inexhaustible source of energy. This is the breeding of fissile U^{233} or Pu^{239} from the all but inexhaustible uranium and thorium contained in the granitic rocks. In the breeder reactor, more fissile material is created from ordinary uranium or thorium than is burned. The breeding process therefore makes every nucleus of uranium and thorium, not just the rare light isotope of uranium, a potential source of energy. As a consequence, uranium ore, which is too expensive to use if only the U^{235} contained in it is burned, becomes an economical fuel. The resulting multiplication of our nuclear energy potential is enormous: first, by the factor of about 400, which represents the ratio of the number of thorium and U^{238} nuclei found in nature to the number of U^{235} nuclei; and second, by the enormously greater factor of perhaps 10^8, representing the ratio of the total amount of uranium and thorium in the accessible parts of the earth's crust to the amount of cheap uranium ore. The total energy content of the residual uranium and thorium in the accessible granites is fantastic — of the same order as the total energy content of the deuterium in the sea. Thus a cheap, practical breeder would provide a permanent, essentially inexhaustible source of energy just as much as would controlled fusion. Moreover, since low-grade deposits of uranium and thorium are ubiquitous, cheap energy eventually would be available in every portion of the globe.

Of course, we would not be driven to "burning the rocks" — that is, using the residual 10 or so parts per million of thorium and uranium in granite — for many, many years. There are vast amounts of uranium and thorium in the rocks at concentrations of 50 or so parts per million. For example, the Conway granites in New Hampshire contain some 30 million tons of thorium at an average concentration of about 50 parts per million. K. B. Brown and his group at Oak Ridge have developed methods of extracting this material at a cost estimated to be only about $35/pound. Even at this price, which is about eight times the current cost of uranium, the burnup cost in a breeder reactor would be less than .02 mill/kwhe. Since, at Harrison Brown's ultimate energy budget, one is burning about 40 tons of fissile material per day, the 30 million tons of thorium contained in the Conway granites alone would last for a very long time indeed — say, a couple of thousand years! Moreover, the mining operation required to supply the world with 40 tons of thorium or uranium every day would not be unreasonable. If this material were supplied by the Conway granites, only one million tons of rock would have to be mined and processed each day. This is but one fourth of the world's 1952 daily production of coal and lignite.[16] The whole mining operation required to sustain the ultimate energy economy would be smaller than the mining operation that now sustains the much smaller, fossil-fuel-based, world energy economy! But these extraordinary possibilities rest on the development of a successful breeder reactor. It is for this reason that I view the development of a practical breeder to be one of the most important technological jobs facing mankind.

Fortunately the technical outlook for a successful, economical breeder reactor is good, even though most of the world's effort in nuclear energy has not gone toward developing breeders. Five experimental breeder reactors have been built, three in the United States, one in the United Kingdom, and

[16]UN Department of Economic and Social Affairs, "World Energy Requirements in 1975 and 2000," *Proceedings of the International Conference on the Peaceful Uses of Atomic Energy 1*, 3, United Nations, New York (1956).

one in the Soviet Union. Several more are scheduled to go into operation within the next half dozen years, among the largest being a 250 Mwe installation on the Caspian Sea that will energize a desalting plant as well as produce electricity. One of the most successful breeder reactors so far has been the one built at Dounreay, Scotland. This machine has operated well at 60 Mw of heat. Its performance has given the United Kingdom Atomic Energy Authority confidence to plan a much larger breeder. In the United States, the nuclear energy effort will probably shift more and more to the development of breeder reactors, and I am confident that a breeder reactor that produces electricity economically will be operating within ten to fifteen years. I believe this achievement would have to be ranked as of extraordinary importance in the history of mankind, only a little less important than the discovery of fission. It certainly would be as important as would the achievement of controlled fusion.

Some among my readers will accuse me of exaggerating when I predict a resolution through nuclear breeding of the competition between population and resources, a resolution that one hopes will give us at least some of the time we need to learn how to control our population permanently. Yet I believe I am not unduly optimistic. I do not, for example, have to invoke nuclear fusion as have others who have speculated on these matters. I have based my judgment on a technology — nuclear breeder reactors — that is really close at hand. The new age of energy is here, and the extravagant claims made for nuclear energy when it was discovered are really coming to pass.

The Revolution in Information

Science seems to be coming through, again in the nick of time, with ways of dealing with the second Malthusian dilemma: the increase in complexity, that is, the imbalance between the individual's capacity to understand and the proliferation of his semantic environment. The underlying scientific achievement — what the late Norbert Wiener called the cybernetic revolution — has many aspects: automation, digital computation, efficient communication, and, perhaps most important, identification of information as an underlying issue in the science of biology. My knowledge of information technology is inadequate, and so I shall content myself with describing a few of the developments in the handling of information with which I am familiar. I shall then examine what we might expect of these developments in helping to resolve the second Malthusian dilemma.

What is happening in automation and digital computation astounds and astonishes. I have still not quite recovered from SKETCHPAD, a device for engineering drafting demonstrated to me at M.I.T.'s Lincoln Laboratory. Imagine a cathode-ray screen, like a TV tube, on which one can "draw" lines with a rather standard device called a "light" pen. Now suppose you are an engineer designing a bridge. You draw, freehand, the trusses, but as you draw, your slightly imperfect lines are replaced by perfect, straight lines, each matched as nearly as possible to the freehand lines. You now insert with a typewriter a number designating a weight at a certain point on the bridge. Immediately the stresses in every truss and member of the bridge appear! SKETCHPAD is not confined to bridges; it can sketch, and compute, electronic circuits, or linkages, or for that matter, a girl's face. It can twirl the linkage so that the engineer can better visualize its working, or it can wink the girl's eyes to amuse the engineer.

This is not all. Already computers are being designed, and components developed, with memories of 10^8 words and nanosecond access time. A single such computer, with satellites spread throughout a research establishment, could simultaneously make out the payroll, operate an experimental reactor, and calculate relativistic wave functions. It gives one

25

the eerie feeling that E. M. Forster's machine is practically here, and raises the specter, as did Forster, of what happens when the machine stops.[17]

The resolution of Malthus' dilemma of energy imbalance offered by the discovery of nuclear fission is relatively clear. Moreover, the cybernetic revolution also helps resolve the energy imbalance. Mass production and automation have greatly increased the number of commodities that we can produce and have enabled us to convert our available energy more efficiently into things we need. Unfortunately, the solution to the second Malthusian dilemma, the imbalance between the proliferating semantic environment and the individual's semantic mechanism, is less clear, partly because the technology of information is younger, and partly because the problem of the proliferating semantic environment is more complex.

For the information revolution adds to the proliferation of the semantic environment at the same time it helps us cope with it. The automatic telephone system complicates the life of the housewife who may now spend several hours each day speaking with her friends. Yet, as I have said earlier, only because the system has become automated has the telephone system as a whole remained viable. Without automation we could not have the complexity that automation itself keeps under control. Though automation displaces workers, many of the jobs from which they are displaced (such as the operation of telephones) are jobs that, without automation, the workers could not do.

One consequence of our proliferating semantic environment is a trend toward specialization. In science the fragmentation caused by the scientific information "crisis" has evoked much attention from government and from the community of scientists, and I shall enlarge on this matter in

[17]"The Machine Stops," *The Collected Tales of E. M. Forster,* 144–197, Alfred A. Knopf, New York (1947).

Part II. The specialization that has afflicted science almost surely must come to the rest of society, if it has not already come. We, each of us, will be able to understand a smaller and smaller fraction of our semantic environment, and in this sense our social organism must fragment. As I shall describe later, one response of science to this specialization has been the emergence of a hierarchy of scientific generalists who spend their time reviewing and compacting literature for their specialist colleagues. Such generalists may have counterparts in society generally. Our lawmakers, in a way, are generalists, as are our newspaper people. And, just as our scientific generalists have trouble keeping up with the details of science, so our lawmakers and newspapermen have trouble keeping up with the details of the society. A U.S. senator from California today represents seventy-five times as many constituents as did one in 1850. He is a generalist of much higher order than was his early predecessor.

Can the technology of information help the generalist maintain sensitive touch with the details of our society? Already it has done much; our central government would be unthinkable without the telephone and the airplane. We begin dimly to see new ways in which the computers with enormous memories might serve the generalist. Computer science today is barely twenty years old. It is not entirely science fiction to imagine, say, a central computer with a memory of 10^9 words, shared by congressmen and connected to satellite computers spread among the constituency. Should a congressman want to ascertain his constituents' views on a subject, he could canvass them ever so much more rapidly and completely than is now possible. Moreover, he could ask complicated questions, and the computer, if properly programmed, could seek out those elements of the answers that would really help him in making up his mind on a crucial issue.

Recently we have come to realize that information is a central concern of the biological sciences. This may have even more effect on our control of the semantic environment than will the computers. Evidence begins to accumulate that the human brain itself may have certain elements that resemble a computer, and that either RNA or some protein may be the essential memory element. J. V. McConnell and his co-workers[18] at Michigan claim that flatworms fed RNA extracted from other flatworms trained to traverse a maze are themselves able to learn the maze better than the controls (although this claim has been challenged by other workers, notably M. Calvin); H. Hydén and E. Egyházi[19] in Sweden claim that the chemical composition of RNA in the Deiters' cells of rats trained to do a balancing act is affected by the training; and D. E. Cameron and L. Solyom[20] in the United States claim that the memories of persons suffering from cerebral arteriosclerosis are significantly improved when RNA is added to their diet. It is too early to say that these findings will be sustained, but it is hardly idle to speculate that many of the mechanisms of the brain will be elucidated, say within a generation, and that from this may come ways of improving the efficiency of our own brains. No matter how good our computers become, human brains finally must monitor their output, must inject the quality of imagination denied to the computer. I feel a little more comfortable about the 10^9 word computers since I see a hope that science might help us improve the working of our own brain-computer, and thereby enable us more effectively to monitor the information robots. We may learn to redress the imbalance between the semantic environment and our individual semantic mechanism not only by using very large computers more cleverly but, perhaps, by making ourselves cleverer.

[18]"Memory Transfer through Cannibalism in Planarians," *Journal of Neuropsychiatry 3,* Supplement 1, S–42–48 (August 1962).

[19]"Nuclear RNA Changes of Nerve Cells During a Learning Experiment in Rats," *Proc. Natl. Acad. Sci. U.S. 48,* 1366–1373 (1962).

[20]"Effects of Ribonucleic Acid on Memory," *Geriatrics 16,* 74–81 (1961).

The Tainted Revolutions: The Applied Scientists' Responsibilities

I have painted an optimistic picture of science's capabilities for resolving Malthus' two great dilemmas. This optimism must be tempered, however, because the solutions offered by science are imperfect; in solving these problems, science creates others. The solutions offered by the two great scientific revolutions centering around energy and information are tainted.

The most obvious taint is the bomb. Nuclear explosives, together with our clever methods of delivery, have given man a relatively easy way to destroy most of what he has. At this stage in the thermonuclear era, one no longer argues about whether the thermonuclear weapon is a blessing in disguise or an unmitigated catastrophe. One simply states to which camp he belongs; I belong to the optimistic camp. I attribute to the bomb the role of peacemaker. The peace we have, tenuous and incomplete as it is, is infinitely better than large-scale war. The simple, unsophisticated notion that the bomb-deterrent has bought us the time we need to get used to the idea of settling international squabbles without large-scale war seems to me to be nearer the actual situation than are any of the more sophisticated views, all of which tend to underestimate the strength of man's instinct for self-preservation.

And, indeed, I believe the energy revolution does have the possibility of helping to stabilize the bomb-imposed, unstable equilibrium. Residual uranium and thorium are available in all the granitic rocks everywhere on earth. When breeder reactors have been developed, every nation, large or small, that can put together the capital to buy the necessary reactors can have abundant and cheap energy. From these central energy sources can flow water from the sea, metals, even liquid fuel. Thus eventually the difference between have and have-not nations, insofar as these differences are based on

29

disparity in natural endowment of raw materials, ought to diminish. And is it not at least plausible that a world no longer beset with widespread hunger and privation, a world afraid to use its nuclear weapons, would be largely a peaceful world?

How this might come about is suggested to me by the effect of nuclear desalting technology on the Middle East. Before desalting technology was recognized as being available to Israel, destruction of the Jordan River Project made a kind of sense to Arab nations bent on destroying Israel. But with desalted water available at about the same price as water from the Jordan system, such action loses much of its point. I would therefore be rash enough to predict that before the century is out, water as such will no longer be a basis for rivalry between Arab and Israeli, and that the disappearance of this source of conflict will eventually lead to improvement of the political climate in the Middle East.

A second taint is the one exemplified in the dramatic writings of the late Rachel Carson: the increasingly serious physical insults to the biosphere imposed by our industrial civilization. Miss Carson spoke only of insecticides, which are needed to help us grow enough food yet which poison our biological environment. But the Rachel Carson problem is only one example of the contamination of our environment that seems to accompany each of our attempts to reduce the imbalance between resources and population. The TVA's Kingston Steam Plant, rated at 1.6×10^6 kilowatts, emits about 400 tons of SO_2 per day into the atmosphere, as well as appreciable amounts of radium. The nuclear reactors I have described create toxic radioactive wastes. Our automobiles help create smog. The whole environment is assaulted by civilization's garbage, and unless curbed, these assaults finally reach the biological world. Before any optimistic view of what science can do to control the Malthusian dilemma is to be taken seriously, one must demonstrate that these taints can be avoided or otherwise dealt with.

I find reason to be hopeful on two accounts: first, we shall learn how to remove the physical insults to the biosphere, and second, we shall learn how to correct the biological damage such insults may cause. With respect to removing the insults, I mention, for example, the very real possibility of economical, pollution-free, electric automobiles. George Hoffman,[21] formerly of the RAND Corporation, has pointed out that the zinc-air battery, one of several types now under development, could provide an ordinary automobile with a range of about 200 miles and a top speed of 95 miles per hour. D. Friedman[22] has recently described a lithium-chlorine electrochemical engine with a specific energy (watt hours per pound) very close to that of a gasoline engine. Its fuel cost, if electricity for recharging were available at 6 mills/kwh, would be competitive with gasoline at 10 cents per gallon! Hoffman is distinctly optimistic about cheap batteries becoming available, and I believe big nuclear reactors eventually will provide electricity to the consumer at less than the 6 mills/kwhe needed to make the electric automobile economical.

Another insult about which much has been said publicly is the possibility of contaminating the environment with the radioactive wastes from large reactors. But routine and safe disposal of radioactive wastes has proved to be simpler than had originally been expected. For example, at Oak Ridge radioactive wastes mixed with cement are being pumped 1000 feet into the ground, there to set permanently along fracture planes between beds of rock. As nearly as geologists can determine, these sheets of radioactive concrete will remain completely out of contact with the biosphere until long after

[21]G. A. Hoffman, "The Electric Automobile — An Example of Vehicle Systems Design," Report MR–54, University of California, Los Angeles (December 1965).

[22]D. Friedman, "The Correlative Advantages of Lunar and Terrestrial Vehicle and Power Train Research," Society of Automotive Engineers, Automotive Engineering Congress, Detroit, Michigan (January 1966).

their radioactivity has decayed. And there are many other feasible schemes for disposing of radioactive materials from reactors, safely and permanently — for example, in unused salt mines or in specially built concrete vaults.

On the matter of radioactive hazard from an inadvertent runaway of a large reactor, such an occurrence can hardly be completely ruled out just as one cannot rule out the possibility of a jet airliner crashing into Yankee Stadium at the height of a World Series game. But major advances have been made in the engineering of "containment" shells, the airtight domes that house nuclear reactors. The general idea is to enclose one containment shell by a second shell, and to keep the space between the two shells below atmospheric pressure. This gas space is continually monitored, so that if any radio-activity appears in the space, it can be handled safely before it escapes to the outside. Generally I am extremely optimistic about dealing with our nuclear garbage, so much so that I believe nuclear plants will displace fossil fuel plants not only because they are cheaper but also because they are cleaner. This seems to have happened in Dade County, Florida. The Florida Power and Light Company, in announcing its decision to build two 750 Mwe pressurized water reactors at Turkey Point, near Miami, explained that only with nuclear plants could the utility meet the stringent requirements imposed by local ordinances regulating the contamination of the atmosphere by power plants.[23]

But there will always be residual physical insults to the biosphere. Is it likely that biologists will learn how to cope with such unfortunate sequelae of exposure to toxic chemicals or radiation as leukemia or gene mutation? Here one can only speculate. The recent discovery of viruses in chemical- and radiation-induced leukemias,[24] and the finding of ways to

[23]McGregor Smith, et al., "Nuclear Power," A Panel Discussion by Utility and Government Experts, Southern Interstate Nuclear Board, Oak Ridge, Tennessee (January 1966).

[24]A. H. Upton, et al., "Observations on Viral, Chemical, and

confer immunity against leukemogenic viruses in experimental mice,[25] are too striking to allow anything but optimism. Many workers in the field believe that the leukemias, now the least tractable of the cancers,[26] ought to be the first curable cancer. Should this take place, science will have removed·one of the taints associated with science's solution to Malthus' first dilemma. As for mutagenesis, recent work suggests that,[27] by inspecting prospective parents' chromosomes, pathologists of the future might identify aberrations that would lead to some birth defects. This work is barely beginning, but its possible implications are very exciting.

The huge size which seems to be required of nuclear reactors if they are to be as cheap as I have postulated is an obvious imperfection in the solution offered by nuclear energy to Malthus' first dilemma. If, in order to produce energy cheaply, a nuclear reactor must be much larger than can be accommodated by existing economic and social organizations, then, unless these organizations are merged and enlarged, we shall have to forgo the economic advantage of bigness. This point is an extension of one made by John von Neumann[28] in 1955. Von Neumann, concerned mostly with the H-bomb and with weather modification, pointed out that the geographic impacts of these technologies are so vast as to have rendered

Radiation-Induced Myeloid and Lymphoid Leukemias in RF Mice," *Journal of the National Cancer Institute* (in press, 1966).

[25]Mary Alexander Fink and Frank J. Rauscher, "Immune Reactions to a Murine Leukemia Virus. I. Induction of Immunity to Infection with Virus in the Natural Host," *Journal of the National Cancer Institute 32*, 1075–1082 (1964).

[26]K. M. Endicott, *Hearings before a Subcommittee of the Committee on Appropriations,* Eighty-Ninth Congress, Department of Health, Education, and Welfare, Part 3, National Institutes of Health, pp. 324–402, U.S. Government Printing Office, Washington, D.C. (1965).

[27]M. Bender, private communication. Also, Robert S. Ledley and Frank H. Ruddle, "Chromosome Analysis by Computer," *Scientific American 214,* 40–46 (April 1966).

[28]"Can We Survive Technology?", *Fortune 51,* 106–108 (June 1955).

obsolete our fragmented political and geographic entities. He suggested that unless the world accommodated to this characteristic of modern technology by closer cooperation between political units, if not reorganization into much larger units, the world could not long survive. Von Neumann was concerned mainly with the impact of military gigantism on our political organizations; I am concerned with the impact of energy gigantism on our economic organization. I believe in both cases the influence of the technology is toward a merging and melding of the relevant political or economic units.

Actually, the economical size of nuclear reactors no longer seems as enormous as it did even in 1960. At that time, a power plant of 500 Mwe was very large indeed. Today, as I have already said, there are a half dozen power plants under construction in the United States with capacities of 1000 Mwe. Moreover, as the over-all size of an electrical system increases, the size of each individual unit on the system tends to increase roughly in the same proportion. A single unit producing 2000 Mwe as part of a station generating 6000 Mwe hardly seems so bizarre; within a decade such plants very probably will be built.

As the size of each unit increases, the penalty that society must pay for technological error increases. This was illustrated dramatically during the 1965 power failure in the Northeast. Here a failure in a huge, interconnected electrical grid system caused considerable social harm, not to speak of loss of property and possibly life. Obviously the scientists and engineers who design such devices bear a heavy social responsibility — heavier, say, than that borne by an architect who designs a small dwelling. My own impression is that the nuclear and electrical engineers will respond adequately to this responsibility by engineering more safety and reliability into their reactors and by analyzing system connections so that a failure in one part of the system will not cause failure in the entire system. The various parts of interconnected electrical grids will probably be tied together more rationally;

at least, the strongly interconnected electrical grids such as the TVA system seem to be more resistant to total failure than are the weakly interconnected systems composed of many independently operated utilities such as were involved in the Northeast incident.

Ordinarily in discussing this aspect of the social responsibility of the scientist, we stress the *recognizable* dangers to society that result if the scientist errs. The tendency then is to put pressure on the technologist or scientist not to try his new schemes because of their evident danger. This is the force of the argument with respect to insecticides, or with respect to the hazards of radioactivity, or the danger of catastrophic collapse of an electrical system. But there is an obvious other side to the story: the *inevitable* catastrophe that society faces because of Malthus' first dilemma, if science and technology do nothing. As I have tried to demonstrate, cheap and abundant nuclear energy is no longer a luxury; it will eventually be a necessity for maintenance of the human condition.

Thus a central social responsibility of the scientist and technologist is to remove the taints, the imperfections inherent in the big technologies needed for mankind's ultimate survival. The task will require reactor and electrical engineers who can reduce the probability of accident to the vanishing point; it will require sanitary engineers and ecologists and chemical engineers who effectively cope with the noxious effluents; and finally, it will require biologists and medical researchers who seek ways of mitigating the biological effects of whatever residual contamination of the biosphere is inevitable.

The Tainted Revolutions: The Humanists' Responsibilities

I suppose I am less hopeful, if not less clear, about the taints associated with the cybernetic revolution — science's contribution to the resolution of the second Malthusian dilemma. Automation and computers, as well as abundant energy, lead to more leisure and to more boredom, and these are taints just as surely as are smog and radioactivity. In earlier times, man's primary concern was economic; making a living was a full-time job. As John Galbraith puts it, ". . . for those who are poor, nothing is so important as their poverty and nothing is so important as its mitigation. . . . And since for nearly all time nearly all people have lived under the threat of economic privation, men of all temperament and views have stressed the controlling and permanent influence of economic need on social attitude."[29] But the cybernetic and energy revolutions suggest that we shall have to modify this observation and say, "For those who are rich, and have leisure, and are bored, nothing is so important as boredom, and nothing is so important as its mitigation."

For the problem of boredom per se, science can supply its brand of antidote. Science, as one of man's supreme intellectual achievements, shares many of the attributes of the arts. And science practiced widely as high culture, as a means of filling empty lives, is surely desirable; but it seems more likely in the immediate future that the arts and the humanistic studies must continue to play the larger role in filling the vacuum created by our taint of too much leisure. I can therefore see the social responsibility of the humanist as being analogous to the social responsibility of the scientist: the scientist primarily to undo the physical taints of the new revolutions, and the humanist to undo the moral taints of the new revolutions.

I use the word "moral" advisedly, for boredom is the lesser of the psychological evils stemming from our new technology. Of greater concern is the "meaninglessness" of human life, which has become a preoccupation of our modern theologians, notably Paul Tillich. To previous generations survival was so arduous that in itself it gave a certain purpose to life. Few men had the time, or even the extra physical energy, to con-

[29]John Kenneth Galbraith, "Economics and the Quality of Life," *Science 145*, 117–123 (July 10, 1964).

cern themselves with life's larger meaning and purpose; to those who worried about the matter, religion was an adequate answer. With our new leisure, as well as our new knowledge, the ultimate questions of meaning and purpose can no longer be submerged because we are too busy. We are not busy, and the historic central practical purpose of human life — economic survival — is no longer sufficient to sustain us.

To reinject meaning and purpose into our lives, all of us must turn to those who traditionally have carried this responsibility, the humanists. How they shall do this I, a scientific administrator, can hardly suggest. Yet do it they must. We scientists, even as we set about correcting the physical defects of our technical revolutions, can only pray that the humanists will supply those deeper values which up to now Western man has had no time to cherish, but which in the future he will have too much time to survive without.

II

THE PROBLEMS OF BIG SCIENCE: SCIENTIFIC COMMUNICATION

Science has become big in two different senses. On the one hand, many of the activities of modern science — nuclear physics, or elementary particle physics, or space research — require extremely elaborate equipment and staffs of large teams of professionals; on the other hand, the scientific enterprise, both Little Science and Big Science, has grown explosively and has become very much more complicated. In a sense, this latter development of science is an example of the so-called second Malthusian dilemma I mentioned previously, the dilemma of complexity (in this case, within science) brought on by the extraordinary growth of science.

The emergence of Big Science has created many difficult problems for science itself, for the institutions at which science is pursued, and for the relations between these institutions and the society that supports them. Some aspects of this relation between our science and our society will be the subjects of the last two parts of this volume. Can we allocate resources rationally between competing branches of science? Can we adjudicate rivalries between different scientific institutions, all of which are supported by the same government? Can we distribute scientific support more uniformly to the different regions of our country and still maintain science as a responsible undertaking?

Here, however, I shall be concerned with how the advent

of Big Science and the explosive growth of science are affecting science itself. Can one, for example, really identify a Big Science syndrome? The research professor in a simpler day concerned himself with the substance of his science, both in research and teaching. Now, through no fault of his own, he must concern himself with many other matters. To do his research, he has to manage, even in Little Science, fairly large sums of government money. He must write justifications for his grants; he must serve on committees that recommend who should receive support, who should not; he must travel to Washington either to advise a government agency or to cajole a reluctant contract administrator. In short, the research professor must be an operator as well as a scientist.

To turn to a fairly obvious point, to what extent has the expense and massiveness of Big Science reduced the flexibility of science? A big research reactor or accelerator is very costly. Its operation keeps people employed and provides a convenient means for turning out scientific papers. To what extent has science suffered because scientists are understandably reluctant to junk a very expensive piece of apparatus, even though the logical force of scientific development would send them in other directions that do not use this particular apparatus? Or, a variant to this question, to what extent are the directions of science influenced by the availability of money rather than the interest and promise of a field?

It would be hard to prove that American science, as a whole, has suffered because its successful practitioners have often been caught in this whirlwind, or because we are reluctant to shut down expensive but unproductive pieces of equipment. As judged by the Nobel Prizes we have won since the war, and the general esteem in which American basic science is held throughout the world, one must conclude that the over-all picture is a healthy one. And yet there are some foreign critics, like Fred Hoyle, who claim to see evidence that the tight intellectual discipline necessary for science is being loosened. Mark Kac[1] has pointed out in speaking of mathematics that Big Science has become democratic science: many more people are involved in it than in previous years. Science has become a way of making a living as well as a way of life. To the extent that this means we have more practitioners of science whose individual dedication is less intense than it was in former times, I would agree that the intellectual discipline of science has been weakened by its growth. No one can say whether this means that we shall have relatively fewer revolutionary advances — breaking of paradigms, as Kuhn[2] puts it — simply because our geniuses are surrounded by more blockheads than before.

These matters that I have touched upon have a sociological connotation. How a scientist spends his time, and whether he now performs less efficiently than he did twenty years ago, are questions probably better answered by the methods of social science than by speculation or personal experience. There are, however, other more philosophic questions concerning the explosive growth of science that are hardly susceptible to scientific study. These have to do with how the intrinsic character of science changes in response to the expansion of science.

The scientific activity that is at once most affected by the expansion of science, and that is perhaps most directly a measure of this expansion, is the scientific communication system. The Malthusian second dilemma — that the communication system grows faster than do the number of communicators — is nowhere better illustrated than in the extraordinary proliferation of our system of scientific communication. In some respects the crisis in scientific communication epitomizes many of the contradictions and

[1]"Mathematics: Its Trends and Its Tensions," Occasional Paper No. 10, 4, The Rockefeller Institute Press, New York (1961).
[2]Thomas S. Kuhn, *The Structure of Scientific Revolutions,* The University of Chicago Press, Chicago, Illinois (1962).

41

limitations of Big Science; the problems of science's system of communication can hardly be separated from the problems of Big Science itself.[3] In writing of the problems of Big Science I shall therefore be largely concerned with the crisis in scientific communication.

The Limits of Science

Eugene P. Wigner, some fifteen years ago in a beautiful little essay called "The Limits of Science,"[4] was one of the first to ask how the n^2 singularity of Malthus' second dilemma would eventually place limits on science. Most important, there is the well-recognized fragmentation of science; each scientist narrows his field of concern, always accommodating his interests to the limitations of his own personal communication system. A physicist who grew up when quantum mechanics was invented at one time knew the entire field. As time goes on he reluctantly and sometimes bitterly must relinquish such all-encompassing grasp. As an older physicist he must confine himself to nuclear theory or to dispersion theory.

I suppose the most obvious danger lurking in the fragmentation of science can be seen by reflecting on the historical fact that the various branches of science have always fertilized each other. Much of modern chemistry would be unthinkable today without the insights provided by quantum physics. Large parts of biology depend upon the tools of the physical sciences and their conceptual insights and frameworks. If the prime purpose in pursuing science is the edification of the individual scientist, then fragmentation causes no great loss: a scientist can get almost as much satisfaction working in a

[3]Many of my views are drawn from the President's Science Advisory Committee's report "Science, Government, and Information," U.S. Government Printing Office, Washington, D.C. (January, 1963).
[4]*Proc. Am. Phil. Soc. 94,* 422–427 (1950).

very narrow specialty appreciated by only a few of his colleagues as in working on a broader canvas. But if the prime
purpose of science is to learn as efficiently as possible as much
as possible about the world, then any breakdown in communication between the sciences or between neighboring
branches within a science is, if not a calamity, certainly a
cause for deep concern.

Growth and fragmentation impair the efficiency of science
by forcing science to become a team activity, because a single
knowledgeable mind is in many ways a more efficient instrument than is a collection of minds that possess an equal total
sum of relevant knowledge. The act of scientific creation, no
less than any intellectual creation, is largely an individual act.
It is done better when all relevant facts are immediately available to be synthesized on the spot by a single would-be
creator. Of course, the sum total of a team's knowledge generally exceeds the knowledge of any member of the team, and
so a team for certain kinds of science, the kinds that do not
require the most original insights, will be more effective than
the individual. Yet, I simply cannot imagine the theory of
relativity, or Dirac's equation, coming out of the teams that
nowadays are so characteristic of Big Science. Even in team
science, the dominant member of the team is the one with
the broadest command of the various specialties represented
by the other members of the team. Thus, as science fragments
and becomes a team activity, it probably becomes less efficient in the sense that more people and more money are
needed per unit of scientific output. It is still true that a
competent scientist with all the requisite facts at his fingertips
will usually get a result more quickly and more cheaply than
three lesser scientists who must pool their knowledge.

There is a deeper danger lurking in the fragmentation of
science. If the separate branches cease to communicate with
each other, what prevents their eventually becoming incon-
sistent with each other? That this is not an idle possibility

was pointed out by the American geologist M. King Hubbert.[5] Apparently, hydrologists had for many years used an equation to describe the flow of incompressible fluids that left out an important term; hydrodynamicists, of course, knew the right equation, but because hydrologists did not speak much with hydrodynamicists, this point in theoretical hydrology contradicted the laws of hydrodynamics, and no one was very alarmed about the situation.

Wigner also foresaw a second trend in science's response to its communication problem: that, as science fragments, it seeks to reintegrate itself by moving to a higher level of abstraction. Quantum mechanics implies the details of the Balmer series in hydrogen; topology now unifies certain aspects of nonlinear differential equations; the Watson-Crick model unifies many parts of biology. But the unification is at a higher level of abstraction. Knowing the relativistic Dirac equation is not the same as sensing and feeling the Balmer lines showing darkly in the sun's spectrum.

In some ways this tendency of science to achieve unification by shifting to a higher level of abstraction poses an all but insoluble difficulty. This is well illustrated by mathematics. Here topology, the study of properties of objects that remain invariant under any transformation that does not tear the object, has become one of the great unifying forces for much of mathematics. For example, the topologist now can classify and order the solutions to large classes of differential equations; he can deduce theorems in the calculus of variations; he can even unify diverse points of view in number theory. But in so doing, he loses something. The metric aspects of the solution to a differential equation do remain important — if not to the topologist, then certainly to one who applies differential equations either to other parts of mathematics or to other sciences. Thus knowing *in principle* is not the same as knowing. Generally as one moves to a higher level of abstrac-

[5]"Are We Retrogressing in Science?", *Science 139*, 884–890 (March 8, 1963).

tion, one omits something, either because applying a complete theory to a detailed situation would go beyond our mathematics (we cannot deduce the energy levels of U^{IV} from quantum theory) or because our original theory omits something initially (we have many models of the nucleus, but each model is inadequate in some respect). Thus quantum mechanics implies all the properties of copper sulfate; but it would be difficult indeed to deduce the blue color of copper sulfate from quantum mechanics. And yet copper sulfate *is* blue, and insofar as science is a description of the world, our science is imperfect if it ignores the blueness of copper sulfate. Although the trend toward unification through abstraction is probably inevitable as science grows, it is well to remember that we pay in loss of resolution for the broadened viewpoint we gain as we become more abstract.

We also pay in another way. As a theory grows more abstract and more general, it usually becomes more difficult. It uses a more specialized terminology; it requires more abstruse and recondite concepts for its understanding. As a practical matter, this means that the number of people who can understand the theory tends to decrease. Often this is temporary: the theory of relativity, viewed as being impossibly difficult at the time it was proposed, is now common knowledge to every physicist. Still I believe there are limits to the degree to which general theories become fully assimilated, if only because keeping up with the general theory itself becomes a full-time job. Thus in a practical sense, mathematics, at the same time it is becoming conceptually more unified, is fragmenting at this higher level of abstraction simply because those whose fields are receiving the benefits of the unification are incapable of communicating with their unifiers.

Finally, the expansion of science accentuates the changeability of scientific fashions. There is always a limit to the number of topics that the entire scientific enterprise can actively pursue. What happens to the older topics when new ones arise to displace them? Sometimes a field becomes obsolete because all the problems are solved; this is largely true of classical thermodynamics or Newtonian mechanics. The field then becomes unfashionable but retains its usefulness in other branches of science. Sometimes fields of science become obsolete not because the field has been resolved, but rather because no one knows how to make progress and scientists lose interest. To an extent this was the case with superconductivity before the isotope effect was discovered: this discovery put the theorists on the right track and regenerated intense interest in superconductivity. Physical optics was regarded as unfashionable among physicists until the laser was invented, at which time physical optics was vigorously reborn. Very typically, a field that was once fashionable eventually ceases to command the interest of the scientists in that field and becomes the concern of scientists in another field. Nuclear chemistry is a good example of this trend: it began as nuclear physics, was taken over by chemists, and now, insofar as nuclear properties of radionuclides are important for technology, parts of nuclear chemistry are being taken over by engineers. This tendency for fashions in science to come and go greatly complicates the teaching of science. For, as science proliferates, the discrepancy tends to widen between the older, consolidated body of scientific knowledge and the parts of science that excite the active researcher. The older mathematician may know both the classical calculus of variations and the topological approach to calculus of variation, but to his students he is likely to present only the topological approach. In the process the traditional parts of the subject tend to be lost. We only begin to see this problem, say in mathematics or physics, because there are still a few giants from a previous day who give the impression of commanding most of the field. But it is doubtful that, with the continuing growth of knowledge, future generations will boast a mathematician who knows as large a fraction of the field as von Neumann did or that physics can have another Pauli, who knew essentially all of physics. The preservation of our traditional scientific knowledge in the face of the explosion of scientific information is one of the intrinsic limits to the growth of science that is not easily resolved. I shall return to this point in Part IV, where I discuss how this trend affects our scientific educational institutions.

The Systematization of Induction

The two trends toward fragmentation and toward abstraction have led Wigner to suggest that science will undergo a social reorganization. He sees the scientific community being layered into several hierarchies. At the first level are the bench scientists, each of whom works in a rather narrow field and each of whom communicates with other closely related bench scientists. The results of each group of bench scientists are kept under surveillance by the next group of scientists, the group leaders or bosses. These group leaders communicate with each other, and in this respect maintain contact between different groups of bench scientists. In principle, the hierarchy could be extended with groups of group leaders themselves being kept under surveillance by supergroup leaders who again would communicate with each other at a higher level of abstraction than do the group leaders. This proposed hierarchical structure for science corresponds to a separation into different levels of abstraction which is reminiscent of Alfred Korzybski's "Structural Differential."[6] The traditional working scientists are at the bottom rung — each one knows almost everything about almost nothing; as one progresses toward the top of the ladder, the subject matter becomes more abstract until one finally reaches the philosopher at the top who knows almost nothing about almost everything.

In some branches of science the theoretical scientist, who works at a higher level of abstraction than do the bench

[6]*Science and Sanity,* 3rd ed., 403, The International Non-Aristotelian Library Publishing Company, Lakeville, Connecticut (1948).

scientists, has manifestly taken on this job of higher order surveillance. I saw this illustrated a few years ago during a meeting on the chemical basis of mutagenesis. Most of the conference participants were experimental biochemists; they knew each other and knew each other's work. Among the group was a distinguished quantum chemist, who on the one hand seemed to know the experimental data better than did most of the other participants (this is not unnatural since theoretical scientists typically spend much time pouring over experimental data amassed by others) and who on the other hand tried in every instance to fit each piece of data into a broader framework. The quantum chemist maintained contact both with the community of quantum organic chemists and with the community of biochemists. He was an example of the group leader who interprets and correlates the results obtained by the individuals in his group (the bench biochemists), and maintains contacts with other group leaders in neighboring sciences — that is, in quantum chemistry.

The theoretical physicist does this sort of job for physics; to a lesser extent, the theoretical chemist does the job for chemistry; and to an even lesser extent, the theoretical biologist does it for biology. Now the main point of all this is that the theoretical physicist plays an important role in helping to resolve the information crisis in physics. His theories, on the one hand, deepen our understanding of the experimental phenomena; but on the other hand, they provide a simplified framework upon which one can hang broad reaches of experimental data. For example, there are now more than 100 elementary particles — mesons, strange particles, and strange particle resonances — and the number is growing. It is a formidable task for anyone who is not working in elementary particle physics, and even for those who are, just to keep track of the particles, to know the difference between a Ξ^0 and a Σ^-, or to remember the isotopic spin of the K meson. But recent theoretical studies have uncovered striking regularities among these particles. For example, the

Ne'eman—Gell–Mann eightfold way not only deepens our insight into the particles but also enables us to keep them straight in our minds. In this sense the theorist, in correlating and providing a conceptual framework for isolated experimental facts, achieves much the same objective as does the information specialist; he, in effect, compacts the literature much more efficiently than can any abstracting service or review article.

There are many other examples of how the theorist compacts the literature. Perhaps the best example for chemistry is the Mendeleev periodic system and Bohr's Aufbau Prinzip which explains the periodic system in terms of quantum theory. The regularities embodied in Mendeleev's chart immediately make a wide range of chemical data available to the average chemist. Before Mendeleev, the chemist had to remember that the hydroxides of Li, Na, K, Rb, and Cs are strong bases; after Mendeleev he needed only to remember that the bases of the Group I metals are strong, and that Li, Na, K, Rb, and Cs belong to Group I, a fact that he would want to remember for other reasons. In this way Mendeleev compacted the literature tremendously for all subsequent generations of chemists. Bohr's insight represents a further compaction of the data: he gives the key to understanding how to construct the Mendeleev table. Moreover, Bohr, plus the ideas of quantum chemistry, enables one to reconstruct the chemical properties of elements; for example, to predict that the hydroxides of Group I metals are strong bases without having to remember this specific fact. In the hierarchical scheme, Mendeleev would be one level above the bench chemist; Bohr would be at a still higher level.

The theorist as compacter of the literature is imperfect. He is most successful in the simple sciences like physics, less successful in the complicated sciences like chemistry or biology, and hardly at all successful in the supercomplicated behavioral and social sciences. This is only a way of saying that theoretical science has become a well-defined technique only in the simple sciences, and its usefulness, either as a means to achieving deeper insights, or as a way of compacting the literature, decreases as the subject becomes more complicated. Yet the successful theorist in his role as compacter of the literature is so useful that one is tempted to invent devices or approaches that can play, in the sciences too complex to be compacted by theory, the role that the theorist plays in the simple sciences. In other words, can the process of compacting the literature by identification of easily remembered regularities in complicated situations be systematized?

What I am describing here is really the inductive method in science: the correlation of many seemingly disparate facts and the identification of regularity in a sea of diversity. Perhaps by examining some of the great inductive insights of past science, we may be able to learn how to encourage and systematize induction. Possibly the greatest piece of scientific induction was Darwin's formulation of the theory of evolution and of natural selection. Darwin's method was induction at its very best: a painstaking, massive collection of data, a persistent and thoroughgoing study of the data, and a brilliant identification of regularities in the data, the result being the theory of evolution. Once the theory of evolution had been stated, large stretches of biology, taxonomy, and paleontology could be systematized: an impossibly complicated field acquired regular features and, in this sense, became easier to comprehend.

A second example of successful scientific induction is the formulation of the shell model of nuclei by J. H. D. Jensen and Maria G. Mayer, for which these two physicists shared a Nobel Prize in 1963. Our picture of the nucleus was dramatically changed by the observation that clusters of neutrons or protons with the "magic" numbers 20, 28, 50, 82, and 126 are very stable. Nuclei containing this many neutrons or protons play for nuclear structure the same role that the rare gases with 2, 10, 18, 36, and so on, electrons play for atomic structure. Though such regularities were suspected very early

50

by Elsasser, the full implication of these early speculations had to wait until large amounts of data on properties of nuclei had been amassed and correlated. And, in fact, collection of the growing mass of nuclear data became an important occupation for many nuclear physicists; had the data not been collected, it is hard to see how Jensen and Mayer could have formulated the theory of the shell structure of nuclei. Now that a sort of Mendeleev table for nuclei is available, the literature of nuclear science in effect is compacted. For example, one remembers that the capture of neutrons by lead is improbable because lead has a magic number of protons, 82, and all magic nuclei have a small tendency to capture neutrons.

The successful inductions all share the same pattern: the data are amassed; they are systematized; someone worries very hard and very long about them and, with luck, discovers the regularities. In the old days of Darwin, or even Mendeleev, the amount of data that had to be handled was sufficiently small that it was possible for the same person — or a small group of persons — to be bench scientist, amasser of data, correlator of data, and inventor of new syntheses. But now the situation is different, especially in the complicated sciences. Scientists are adding a huge flood of new facts and observations to our existing store of data. A division of labor between those who create or discover the facts and those who sift, absorb, and correlate the facts seems to be inevitable. The division of labor corresponds very nearly to the social layering of science I mentioned earlier.

Science has, in groping and uncertain fashion, begun to invent new institutions that correspond to this social layering of scientists into sifters and interpreters on the one hand and collectors and inventors on the other. The scientists who collect and sift the facts would in our modern terminology make up a "specialized information center," that is, a group of scientists who, in a narrow, well-defined field of science, collect, sift, and interpret information for other workers in the field. The key word here is interpret: the scientists who man such centers not only handle the relevant documents, but they also glean the scientific gems from the documents and try to make sense out of them. The people who run the center must be recognized scientific leaders in their fields of science, for in the process of examining the data and finding new correlations, they create new science. The specialized information center as information broker has received increasingly wide acceptance since the President's Science Advisory Committee report on information was published in 1963. For example, in 1965 the National Bureau of Standards established the National Standards Reference Data System, a loosely articulated collection of information centers. There are now over 400 specialized information centers in various fields of science in the United States, and their number will undoubtedly grow.

I look upon the information center as a helpful element in the systematization of the process of induction in modern, complicated sciences. In the simple sciences like physics which already have a well-defined theoretical structure, the output of the information centers — for example, the magnetic moments of all the odd-neutron nuclides — would be fed to the theoretical scientists, who try to make the generalizations that encompass all the data. In the more complicated sciences, where the role of the theoretical scientist is not as well developed, the output of the information center itself hopefully shows up regularities which were not suspected before the information center organized the raw data.

In a certain sense, the specialized information center acts as "staff" for the working scientists in the field it serves. Every executive has a staff to do his leg work, much of which consists of ferreting out information that the executive is unable to extract because he is too busy. The staff's value lies not so much in finding relevant documents for the executive to read, but rather in distilling from the documents the information the executive needs. Ideally, each working scientist in this day of information explosion would have his own personal staff to provide him with the necessary information. Obviously this is an impossibly expensive luxury, and science seems to be hitting on the next best alternative: providing information centers as "staff" to serve specialized subfields of science. And, just as the executive's staff often provides important new insights because of the staff's intimate knowledge of certain details, so one sees the information centers occasionally providing significant new insights that come only from the marshaling of many bits of data.

A Digression on Technical Writing: A Lost Art

In calling attention to the analogy between the busy scientist's information center and the busy executive's staff, I cannot resist digressing from my main theme to comment on a very usual complaint that one hears from the executive: Why don't people present their ideas in a readable style? I believe that this complaint can also rightly be voiced by the busy scientist: an important block to effective use of written technical material is the dull and unclear way in which so many scientists and engineers express themselves. The path from scientist to written document to depository to retrieval and back to scientist (what is called the "information transfer chain") is not complete until a relevant part of the content of one person's brain cells becomes the property of another's brain cells. Language must be used in this last stage of the chain. Thus, the use of natural language, which has traditionally been the concern of English teachers, should be a matter of concern to those troubled by the scientific information crisis.

Because of my interest in improving the quality of technical writing, I tried a little experiment at Oak Ridge a few summers ago. We happened to have in Oak Ridge a number of college English and humanities teachers who, under the auspices of the Oak Ridge Institute of Nuclear Studies, were studying science for humanists. Six members of the group,

53

headed by Professor W. H. Davenport of Harvey Mudd College, organized "Project Literacy." The purpose of Project Literacy was first to figure out how and when American technical writing became as turgid and unclear as it now is, and second, to learn what could be done about it. Though the project hardly had time to come to definite conclusions, some of what we found was quite striking.

The style of American scientific writing apparently changed around the end of World War I. Before that time the style was personal (with many "I's"); it was figurative (globs of lava falling from a volcano were likened to a sheaf of arrows sprinkling to the ground); it was active; it did not use nouns as modifiers. Around 1920, the style became impersonal, literal, passive, and peppered with strings of nouns modifying other nouns, until by 1963 we find a sentence, "Boration was effected in the water," meaning simply, "We dissolved boric acid in the water."

We have only hunches as to what caused this drastic change and deterioration in style. I think it may in part be connected with the increasing influence of government on our technology.[7] Government institutions are necessarily impersonal and rigid, and their style of communication tends to be correspondingly impersonal, rigid, and unclear. Whatever the cause, I believe that the present manner of expressing ourselves adds significantly to the information problem, and that we must do something about it.

It seems to me that a most urgent matter is to try to decide just what makes for good scientific style. Books have been written on the subject, but I know of little real experiment to decide what style of writing best conveys a message. For example, would a more figurative, discursive language peppered with "I's" actually convey its message more efficiently to a person who tries to scan many abstracts every month

[7] For an illuminating discussion of the influence of our space enterprise on the style of scientific language, see David McNeill, "Speaking of Space," *Science 152*, 875–880 (May 13, 1966).

than does the accepted terse and impersonal style? Many editors of scientific journals automatically cross out the word "I"; is there any evidence that its judicious use reduces the understandability of technical writing? Or again, what really constitutes an informative title? Are there any principles that tell us how frequently meaty words in a title should follow each other?

One might cast such questions aside as being unworthy of professional scientists and engineers. But I insist that they are part of a rational approach to transfer of information. Much of the discussion of language by documentalists today is concerned with how the language looks to the "neural" network within a computing machine. At least as important, and perhaps much more important, is the analysis of how the syntactical structures of a language look to the neural network within the brain. Such an analysis would require psychologists, natural scientists, linguists, and writers with a feeling for style. I would hope that out of such studies would come a technical language that is at once graceful, easy to write, and easy to comprehend, and whose general use would ease that part of the information problem which our present graceless, difficult, and ineffective style has helped to create.

Transfer of Documents[8]

Usually when an expert speaks of information retrieval he means finding the documents which contain relevant information. But document retrieval is only part of information retrieval. Information is not transferred unless what is presented to the human mind is assimilated. Thus the information problem naturally divides into two separate problems: how to identify, index, store, and retrieve documents, and how to transfer the contents of documents to the mind of the user. Most of what I have said so far about the role of the informa-

[8]Much of the remainder of this essay is based on an article, "Scientific Communication," that I wrote for *International Science and Technology*, 65–74 (April 1963).

tion center and the importance of language is concerned with the second, less usual aspect of *information* retrieval; I shall now make a few observations on the matter of *document* retrieval and the related questions of indexing and initial dissemination.

Most of us in thinking about retrieval of documents probably think of the elaborate mechanical systems based on large memory computers. But the big computer only stores information about a document and retrieves it quickly. It does not help with two other equally important steps in the information transfer chain: initial dissemination, and indexing and cataloguing. Retrieval hardware is no better than the indexing system that enables the machine to selectively reach the required documents.

Much thoughtful attention is being given to improving indexing. One interesting new approach is the citation index. All of us are familiar with lists of references at the end of an article which enable the reader to trace backward in time the antecedents of the particular article being perused. Every scientist has used such lists to delve more deeply into the subject he is studying. But reference lists only go backward in time; they give no hint as to the influence a given article has had on the development of the subject after the article appeared in print. The citation index is a list of the articles that, subsequent to the appearance of an original article, refer to or cite that article. It enables one to trace forward in time the same sort of interconnections with the literature that, by means of references, one now traces backward in time. The National Science Foundation has sponsored trials of citation indexing in genetics and in statistics and probability. And the Institute for Scientific Information in Philadelphia now provides citation-indexing services in several fields of science. Citation indexing, particularly in combination with permuted-title indexing, will in time doubtless be widely used. Its use will further alter both the way in which we think of the technical literature and the way we manage it.

Initial dissemination of technical literature has for the last 100 years been largely by way of the technical journal. Various trends seem to be changing the status of the journal. On the one hand, the journal itself has become so bulky, and the number of journals has so grown, that the journals are often not read. On the other hand, private and semiprivate methods of circulation have sprung up, such as the government's informal report system in technology and the preprint system in basic science. Much of the government's original concern with the "information crisis" was prompted by a growing realization that much of the information contained in the flood of government reports was being lost. With respect to preprints, science faces a real danger of reverting to the privacy of the seventeenth century; some biologists I know think this has already happened in molecular biology, where preprints are often circulated only to one's friends.

A possible technical solution to the problem of adequate dissemination of documents is the centralized depository — that is, a central store to which manuscripts are submitted and which on request distributes copies of manuscripts. The central depository would announce its accessions both by title lists and by abstract bulletins. This idea has been proposed innumerable times, and its advantages are impressive. It is extremely fast; it regularizes distribution of preprints; it compacts the circulating literature; it funnels the accumulation from a given field into one place for efficient retrieval. In spite of these advantages, the technical community generally has rejected the central depository, largely, I believe, because scientists are reluctant to give up the prestige that attaches to publication in a technical journal. Yet several new developments may cause a change in the traditional attitude of technical people toward the central depository.

Perhaps most important, the internal information systems of many government agencies, based upon the informal report, are now essentially full-scale working central depositories. About 100,000 reports were issued in 1962 by govern-

57

ment contractors; most of these are stored and kept under bibliographic control in one of the three major depositories: DOD's Defense Documentation Center, AEC's Division of Technical Information Extension, and NASA's Office of Scientific and Technical Information. Improved hardware has been important to the relative success of these government-operated central depositories. Since the collections are fairly small (that is, by the standards of the Library of Congress), the memories of commercially available computers are large enough to handle their entire catalogues. The revolution in reproduction equipment has pretty well solved the problem of quickly making legible copies from microfilm; thus the store itself is compacted, and the user gets good service.

There is a second, less well-defined trend which also points toward the central depository as an ultimate means of disseminating and storing documents. This trend is exemplified by the American Physical Society's practice of dividing its contributions into those that are very timely and those that are more archival. The former are published in abbreviated form with less than a month's delay in *Physical Review Letters*. The latter are published in full in *The Physical Review* with a four- to six-month delay, and abstracts of these archival articles are distributed as part of the *Letters*. *The Physical Review* is becoming a sort of central depository; many physicists read only the abstract bulletin and consult the full article in their library or obtain reprints from the author. I would expect that many other scientific fields would adopt the physicists' approach to initial dissemination, especially since the physicists seem to be so satisfied with it.

The Role of Government

I turn now to our government's role in the handling of scientific information. But first, why has the government shown intense interest in the system of scientific communication? The most obvious reason is that the government, being the largest supporter of research, accepts a responsibility for the general health of the research it supports. Insofar as good scientific communication is a prerequisite for healthy science, and especially, efficient application of science, the government must be concerned about science's communication system. But there is a second, less obvious reason for the government's concern with the health of the scientific communication system. This has to do with the validity of science as a responsible undertaking, largely supported by the public. The "refereed" literature, by means of which scientists criticize each other and maintain the standards of science, is one of the most important means of maintaining science as a responsible undertaking. It should therefore be a matter of great concern to the government that supports science if traditional means of establishing and maintaining the validity of the scientific enterprise are submerged in a torrent of uncontrolled scientific information.

In carrying out this responsibility, the federal government is confronted with two separate though related problems. On the one hand, many federal agencies operate elaborate internal information systems based largely on the contractor-written report; since so much federally financed development work is recorded only in such reports, the government must keep these internal systems in good order. On the other hand, the federal government, as the supporter of three fourths of all research and development in the United States, must also be concerned with the nonfederal technical communication network. The two communication systems interpenetrate; the federal government must therefore be sensitive to the relation between the two networks.

The complexities that face one in trying to understand the federal technical communication system, let alone advising how to improve this system, are appalling. One finds little uniformity in purpose or method of operation among the many information systems of the agencies, nor is it clear that there should be full uniformity. Some internal agency information systems handle only material originating from research paid for by their agency; others, such as AEC's Division of Technical Information Extension, handle material related to nuclear science, regardless of its origin, on the plausible assumption that an advance in, say, British reactor technology would be of interest to a USAEC reactor contractor. Some agencies, such as NIH or NSF, tend to the belief that scientific communication is best done by the individual scientist through his traditional channels; others, such as NASA and AEC, superpose their own internal communication system on the private systems.

The seeming disorder is perhaps better understood if we recognize that the whole technical community is composed of many subcommunities which crisscross and overlap. The two most important classes of subcommunities are those organized around missions and those organized around disciplines. Most of the large government information systems were originally intended to serve a mission-oriented community; most of the major nongovernment information systems were originally discipline-oriented. As the government agencies have grown, their missions have often expanded so as to include such general tasks as "supporting basic research relevant to national defense." Hence, the original rather sharp distinction between the two systems has been irrevocably blurred.

It is impractical to set back the clock to Vannevar Bush's original idea of a National Science Foundation taking care of most of the non-mission-oriented science, and the other agencies, having well-defined missions, supporting only mission-oriented science. We have generally come to the view that diversity of government support in science, even in a single branch of science, is desirable. But, though diversity of support may be advantageous, diversity of information systems, serving the same technical community, surely is not. Take, for example, high-energy physics, a field supported by AEC, DOD, NSF, and NASA. If the information created under the auspices of each of these agencies were separately collected in agency announcement bulletins, abstract journals, or technical reviews, the worker in high-energy physics would have to peruse four sources rather than one. So to speak, source of support is not a valid criterion for bibliographic classification.

To cope with this situation, the PSAC report "Science, Government, and Information" put forth the following suggestion: that the federal government allocate responsibility for information in each field of science and technology it supports to one appropriate agency. Where the field is supported entirely by the agency there is no problem: nuclear reactors are obviously AEC's business; naval vessel design, DOD's. Where the field is supported by many agencies, the agency now carrying the largest share of the responsibility would then assume full responsibility for information activities in the field. In becoming a "delegated agent" for a given field, an agency would assume many responsibilities beyond merely collecting, announcing, and abstracting relevant material. The agency would establish and support specialized information centers; it would support worthy publishing ventures that would otherwise not receive support; and it would generally take active leadership in encouraging better communications in the field both within and without the government. In effect, the agencies, as delegated agents, would become "fairy godmothers" for communication in the field they cover. This is not without logic. Communication is an essential part of research. If an agency sponsors research, it also ought to allocate resources to support communication necessary for effective conduct of that research; at the very least it ought to be assured that, if it does not do so itself, another agency is adequately looking after the matter.

Exactly what organization the delegated agents should adopt would naturally vary from case to case. In all cases the PSAC report recommended that the agency establish a highly placed "focal point of responsibility," that is, a highly placed

61

official, to carry out the agency's broad responsibilities for information activities. The focal point of responsibility would be part of the research and development arm, not of some administrative arm, of the agency.

Any suggestion that the government do more than it is now doing raises the usual specter of creeping bureaucracy. But it is plain that government involvement in scientific communication will grow, just as government involvement in science and technology is growing. The problem is to guide the growth so that user-sensitive nongovernment systems are not swamped by elaborate government systems that are less user-sensitive. The user-sensitivity and the rigidity of scientific standards of the better technical society information systems are precious things; we should not lightly replace such systems with ad hoc systems conjured up by government bureaucrats.[9] I believe the foregoing recommendations can help forestall this danger. In the first place, the focal point of responsibility in each agency is part of the agency's research and development management. Much of the user-insensitivity of government systems results from their being part of administration rather than of research. Second, by delegating responsibility for handling of information generated in mission-oriented fields of research to the corresponding mission-oriented agencies, the government would be represented by technical men who are usually members of the technical community that centers around the agency's mission. The focal points of responsibility in, say, NIH, would presumably be medical research people. They would be expected to speak the languages of both the technical community and the documentalist; when they sponsored information activities, they could be expected to temper any expansionistic predilection of the government with an understanding of what the field really needed. For example, they would be sensitive to the

[9]For a very urgent warning of the dangers lurking in government-controlled systems, see Simon Pasternack, "Is Journal Publication Obsolescent?", *Physics Today* 19, 38–43 (May 1966).

importance of the "refereed" literature in maintaining scientific standards, and they would resist attempts to undermine the strength of this literature by a flood of unrefereed, government-supported reports. Finally, I point out that *support* of an activity by government does not necessarily mean *domination* by the government. Much of what an agency spends for information handling would be spent by contract with professional societies that can be expected to retain much autonomy in their choice of how to spend government money for scientific communication.

The Scientist's Responsibility for Communication

The main point that I want to stress in this brief discussion of science information is that document retrieval, which is largely what automation is concerned with, is not enough. Improvements in scientific communication will require scientific middlemen of various sorts who take up where the machines leave off, and who are concerned with information where the machines are concerned with documents. Where can science find these middlemen?

If our scientific communication system is to remain effective, the technical community will have to accept the idea that handling of technical information is a worthy and integral part of science. Science has long recognized teaching as a valid part of the scientific enterprise; it has been willing to devote a sizable fraction of its resources, and its able people, to teaching, and has accorded high honors to the successful teacher of science (who, generally, is also a successful researcher). In the same spirit, some of the really able practitioners of science must be prepared to devote themselves to the meaningful handling of information, and science must accord them honor commensurate with the importance of their task.

Perhaps the following historical analogy will help persuade technical people that there is considerable personal reward to be gained from devoting more of their efforts to sifting, compacting, and interpreting the literature, and to serving on science's "staff." The ancient civilizations of the Indus Valley, in what is now West Pakistan, depended for their existence upon an elaborate network of irrigation canals. The water flow through these canals was regulated by the gate-keeper; he decided who was to receive water and who was not. As time went on, the gatekeepers of the irrigation systems acquired more and more power; eventually, according to some historians, they became the rulers of the areas through which the water flowed.

This theory of the origin of the ancient kings of southern Asia has an analogy in our modern science and technology. The ideas and data that constitute science are embodied in its literature. The proper flow of the information contained in this literature is necessary for the existence of science in much the same way that the proper flow of irrigation water was necessary for the existence of the Indus Valley civilizations. I find it attractive to imagine that just as the ancient gatekeepers became kings, so as our science and technology grow and the flow of information becomes a crucially difficult problem, those who control information will become a dominant priesthood in the science of the future. I should think this possibility for extending their scientific influence ought to persuade many able people to devote more of their energies to coping with the flood of information.

III

THE CHOICES OF BIG SCIENCE

1. *Criteria for Scientific Choice*[1]

As science grows, its demands on our society's resources grow. It seems inevitable that the demands of science will eventually be limited by what society can allocate to science. We shall then have to make choices of two kinds. We shall have to choose among different, often incommensurable fields of science — between, for example, high-energy physics and oceanography, or between molecular biology and the science of metals. We shall also have to choose among the different institutions — universities, government laboratories, and industry — that receive government support for science. The first choice I call scientific choice; the second, institutional choice. My purpose is to suggest criteria for making scientific choices and to formulate a scale of values which might help establish priorities among scientific fields whose only common characteristic is that they all derive support from the government.

Choices of this sort are made at every level, both in science and in government. The individual scientist must decide what science to do, what not to do; the totality of such judgments

[1]*Minerva I,* 159–171 (Winter 1963); also *Physics Today 17,* 42–48 (March 1964).

makes up his scientific taste. The research director must choose which projects to push, which to kill. The government administrator must not only decide which efforts to support, but he must also decide whether to have a piece of work done in a university, a national laboratory, or an industrial laboratory. The sum of such separate decisions determines our policy as a whole. I shall be concerned mainly with the broadest scientific choices: how should government decide between different branches of basic science? I shall not discuss here the equally important question of how government should allocate its support for basic research among industry, government laboratories, and universities.

Most of us like to be loved; we hate to make choices, since a real choice alienates the party that loses. If one is rich — more accurately, if one is growing richer — choices can be avoided. Every administrator knows that his job is obviously unpleasant only when his budget has been cut. Thus the urgency for making scientific or institutional choices has largely been ignored both in the United States and elsewhere because the science budget has been expanding so rapidly. The United States government spent about \$1 billion in 1950 on research and development, \$8 billion in 1960, and \$15 billion (including space) in 1964.[2]

Though almost all agree that choices will eventually have to be made, some well-informed observers insist that the time for making the choices is far in the future. Their arguments against making explicit choices have several main threads. Perhaps most central is the argument that since we do not make specific choices about anything else, there is no reason why we should make them in science. Since we do not explicitly choose between support for farm prices and support for schools, or between highways and foreign aid, why should

2"Federal Funds for Research, Development, and Other Scientific Activities, Fiscal Years 1964, 1965, and 1966" *XIV*, Surveys of Science Resources Series, National Science Foundation, NSF 65–19, p. 3, U.S. Government Printing Office, Washington, D.C. (1965).

we single out science as the guinea pig for trying to make choices? The total public activity of our society has always resulted from countervailing pressures, exerted by various groups representing professional specialties, or local interests, or concern for the public. The combination that emerges as our federal budget is not arrived at by the systematic application of a set of criteria. Even the highest level of authority (in the United States, the President), who must weigh conflicting requests in the scale of public interest, is limited by the sheer size of the budget, if by nothing else, in the degree to which he can impose an over-all judgment. Because we have always arrived at an allocation by the free play of countervailing pressures does not mean that such free interplay is the best or the only way to make choices. In any case, even if our choices remain largely implicit rather than explicit, they will be more reasonable if persons at every level, representing every pressure group, try to understand the larger issues and try to mitigate sectional self-interest with concern for broader issues. The idea of conflicting and biased claims being adjudicated by an all-knowing supreme tribunal is a myth. It is much better that the choices be decentralized and that they reflect a concern for the larger interest. For this reason alone philosophic debate on the problems of scientific choice should lead to a more rational allocation of our resources.

A second thread in the argument of those who refuse to face the problem of scientific choice is that we waste so much on trivialities — on smoking, on advertising, on gambling — that it is silly to worry about expenditures of the same scale on what is obviously a more useful social objective, the increase of scientific knowledge. A variant of this argument is that with so much unused steel capacity or so many unemployed we cannot rightly argue that we cannot afford a big synchrotron or a large manned space venture.

Against these arguments I would present the following considerations on behalf of a rational scientific policy. At any given instant only a certain fraction of our society's resources goes to science. To insist or imply that the summum bonum of our society is the pursuit of science and therefore that all other activities of the society are secondary to science — that unused capacity in the steel mills should go to Big Science rather than to a large-scale housing program — is a view that might appeal to the scientific community. It is hardly likely to appeal so strongly to the much larger part of society that elects the members of the legislature, and to whom, in all probability, good houses are more important than is good science. Thus, as a practical matter we cannot really evade the problem of scientific choice. If those actively engaged in science do not make choices, they will be made by the Congressional Appropriations Committees and by the Bureau of the Budget, or by corresponding bodies in other governments. Moreover, and perhaps more immediately, even if we are not limited by money, we shall be limited by the availability of truly competent men. I think there is some evidence that our ratio of money to men in science is very high,[3] and that in some parts of science we have gone further more quickly than the number of really competent men can justify.

Choice and Scientific Criticism

Our science and government communities have evolved institutional and other devices for coping with broad issues of scientific choice. The most important institutional device in the United States is the President's Science Advisory Committee, with its panels and its staff in the Office of Science

[3]This belief is based on the fact that between 1950 and 1962 the federal operating budget for science increased nearly tenfold ["Federal Funds for Science X," NSF 61–82, p. 39, U.S. Government Printing Office, Washington, D.C. (1962)], whereas the number of Ph.D.'s increased by less than a factor of two ["Scientific and Technical Manpower Resources," NSF 64–28, p. 141, U.S. Government Printing Office, Washington, D.C. (1964)].

and Technology. This body and its panels help the Bureau of the Budget to decide what is to be supported and what is not to be supported. The panel system, however, suffers from a serious weakness. Panels usually consist of specialized experts who inevitably share the same enthusiasms and passions. To the expert in oceanography or in high-energy physics, nothing seems quite as important as oceanography or high-energy physics. The panel, when recommending a program in a field in which all its members are interested, invariably argues for better treatment of the field: for more money, more people, more training. The panel system is weak insofar as judge, jury, plaintiff, and defendant are usually one and the same.

The panel is able to judge how competently a proposed piece of research is likely to be carried out; its members are all experts and are likely to know the good research workers in the field. But because the panel is composed of experts, with parochial views, the panel is much less able to place the proposal in a broader perspective and to say whether the research proposal is of much interest to the rest of science. We can answer the question "how" within a given frame of reference; it is impossible to answer "why" within the same frame of reference. It would therefore seem that the panel system could be improved if representatives, not only of the field being judged, but also of neighboring fields, sat on every panel judging the merits of a research proposal. A panel judging high-energy physics should have some people from low-energy physics; a panel judging low-energy physics should have some people from nuclear energy; a panel judging nuclear energy should have some people from conventional energy, and so on. I should think that advice from panels so constituted would be tempered by concern for larger issues; in particular, the support of a proposed research project would be viewed from the larger perspective of the relevance of that research to the rest of science.

In addition to using panels or groups (such as the President's Science Advisory Committee) as organizational instruments for making choices, the scientific community has evolved an empirical method for establishing scientific priorities — for deciding what is important in science and what is unimportant. This is the scientific literature.[4] The process of self-criticism, which is integral to the literature of science, is one of the most characteristic features of science. Nonsense is weeded out and held up to ridicule in the literature, whereas what is worthwhile receives much sympathetic attention. This process of self-criticism embodied in the literature, though implicit, is nonetheless real and highly significant. The existence of a healthy, viable, refereed scientific literature in itself helps assure society that the science it supports is valid and deserving of support. This is a most important, though little recognized, social function of the scientific literature.

As an arbiter of scientific taste and validity, scientific literature is deficient in two respects. First, because of the information explosion, the literature is not read nearly as carefully as it used to be. Nonsense is not so generally recognized, and the standards of self-criticism, which are so necessary if the scientific literature is to serve as the arbiter of scientific taste, are inevitably looser than they once were.

Second, the scientific literature in a given field tends to form a closed universe; workers in a field, when they criticize each other, tend to adopt the same unstated assumptions. A referee of a scientific paper asks whether the paper conforms to the rules of the scientific community to which both referee and author belong, not whether the rules themselves are valid. The editors and authors of a journal in a narrowly specialized field are, so to speak, all tainted with the same poison.

Can a true art of scientific criticism be developed — that is, can one properly criticize a field of science beyond the kind of criticism that is inherent in the literature of the field? Mortimer Taube in *Computers and Common Sense*[5] insists

[4]I mentioned this point in Part II, page 59.
[5]Columbia University Press, New York (1961).

70

that such scientific criticism is a useful undertaking. He says that by viewing a field from a somewhat detached point of view, it is possible to criticize a field meaningfully, even to the point of calling the whole activity fraudulent, as he does in the case of non-numerical uses of computers. I happen to believe that Taube does not make a convincing case with respect to certain non-numerical uses of computers, such as language translation. Yet I have sympathy for Dr. Taube's contention — that, with science taking so much public money, we must countenance and even encourage discussion of the relative validity and worthwhileness of the various fields of science supported by society.

Internal Criteria for Choice

I believe that criteria for scientific choice can be identified. In fact, several such criteria already exist; the main task is to make them more explicit. The criteria can be divided into two kinds: internal and external. Internal criteria are generated within the scientific field itself and answer the question, How well is the science done? External criteria are generated outside the scientific field and answer the question, Why pursue this particular science? Though both are important, I think the external criteria are the more important as far as the question of large-scale public support of science is concerned.

Two internal criteria can be easily identified: (1) Is the field ready for exploitation? (2) Are the scientists in the field really competent? Both these questions are answerable only by experts who know intimately the field in question and who know the people personally. These criteria are therefore the ones most often applied when a panel decides on a research grant; in fact, the primary question in deciding whether to provide government support for a scientist is usually, How good is he?

I believe, however, that it is not tenable to base our judg-

71

ments entirely on internal criteria. As I have said, we scientists like to believe that the pursuit of science as such is society's highest good, but this view cannot be taken for granted. For example, we now suffer a shortage of physicians, probably to some extent because many bright young men who would formerly have gone into medical practice now go into biological research.[6] Government support is generally available for postgraduate study leading to the Ph.D. but not for study leading to the medical degree. It is by no means evident that society gains most from more biological research and less medical practice. Society does not *a priori* owe the scientist, even the good scientist, support any more than it owes support to the artist or to the writer or to the musician. Science must seek its support from society on grounds other than that the science is carried out competently and that it is ready for exploitation; scientists cannot expect society to support science because scientists find it an enchanting diversion. Thus, in seeking justification for the support of science, we are led inevitably to consider external criteria for the validity of science, those criteria external to science or to a given field of science.

External Criteria for Choice

Three external criteria can be recognized: technological merit, scientific merit, and social merit. The first is fairly obvious. Once we have decided, one way or another, that a certain technological end is worthwhile, we must support the scientific research necessary to achieve that end. Thus, if we

[6]The relative number of A students in first-year medical school in the United States fell from 40 per cent in 1950 to 13.4 per cent in 1960, according to Walter S. Wiggins, Glen R. Leymaster, C. H. William Ruhe, A. N. Taylor, and Anne Tipner, "Medical Education in the United States," *Journal American Medical Association 178,* 601 (November 11, 1961). A study by the Staff of the Council on Medical Education reported that in 1964, 13.8 per cent of those in first-year medical school were A students — "Medical Education in the United States," *JAMA 194,* 754 (November 15, 1965).

have set out to learn how to make breeder reactors, we must first painstakingly measure the neutron yields of the fissile isotopes as a function of energy of the bombarding neutron. As in all such questions of choice, it is not always so easy to decide the technological relevance of a piece of basic research. The technological usefulness of the laser came after, not before, the principle of optical amplification was discovered; in general, indirect technological or scientific benefits — spin-off — are not uncommon. But it is my belief that such technological bolts from the scientific blue are the exception, not the rule, and that solving a technological problem by waiting for spin-off from an entirely different field is rather overrated. Most programmatic basic research can be related fairly directly to a technological end, at least crudely, if not in detail.

The broader question whether the technological aim itself is worthwhile must be considered again partly from within technology by answering such questions as, Is the technology ripe for exploitation? Are the people any good? It must also be dealt with partly from outside technology by answering the question, Are the social goals that are attained, if the technology succeeds, worthwhile? Many times these questions are difficult to answer, and sometimes they are answered incorrectly. For example, in 1952 the United States launched an effort to control thermonuclear energy on a rather large scale because it was thought that achievement of controlled fusion was much closer than it has proved to be. Nevertheless, despite the fact that we make mistakes, technological aims are customarily scrutinized much more closely than are scientific aims; at least we have more practice discussing technological merit than scientific merit.

The criteria of scientific and social merit are much more difficult: scientific merit, because we have given little thought to defining it in the broadest sense; social merit, because it is difficult to define the values of our society. As I have suggested, the answer to whether a broad field of research has scientific merit cannot be given within the field. The idea that the scientific merit of a field can be judged better from the vantage point of the scientific fields in which it is embedded rather than from the point of view of the field itself is implicit in the following quotation from the late John von Neumann: "As a mathematical discipline travels far from its empirical source, or still more, if it is a second and third generation only indirectly inspired by ideas coming from reality, it is beset with very grave dangers. It becomes more and more pure aestheticizing, more and more purely l'art pour l'art. This need not be bad if the field is surrounded by correlated subjects which still have closer empirical connections or if the discipline is under the influence of men with an exceptionally well developed taste. But there is a grave danger that the subject will develop along the line of least resistance, that the stream, so far from its source, will separate into a multitude of insignificant branches, and that the discipline will become a disorganized mass of details and complexities. In other words, at a great distance from its empirical source, or after much 'abstract' inbreeding, a mathematical subject is in danger of degeneration. At the inception the style is usually classical; when it shows signs of becoming baroque, then the danger signal is up."[7]

I believe there are any number of examples to show that von Neumann's observation about mathematics can be extended to the empirical sciences. Empirical basic sciences which move too far from the neighboring sciences in which they are embedded tend to become baroque — that is, they tend to concern themselves with elaboration of complex details of interest primarily to the experts in the field. Relevance to neighboring fields of science is therefore a valid measure of the scientific merit of a field of basic science. Insofar as our aim is to increase our grasp and understanding of the universe, we must recognize that some areas of basic science

[7]R. B. Heywood (Ed.), *The Works of the Mind,* 196, University of Chicago Press, Chicago, Illinois (1947).

do more to round out the whole picture tnan do others. A field in which lack of knowledge is a bottleneck to the understanding of other fields deserves more support, and should be pushed more urgently, than a field which is isolated from other fields. This is only another way of saying that, ideally, science is a unified structure and scientists, in adding to the structure, ought always to strengthen its unity. Thus, the original motivation for much of high-energy physics is to be sought in its elucidation of low-energy physics; or the strongest and most exciting motivation for measuring the neutron capture cross sections of the elements lies in the elucidation of the cosmic origin of the elements. Moreover, the discoveries that are acknowledged to be the most important scientifically have the quality of bearing strongly on the scientific disciplines around them. For example, the discovery of x rays was important partly because it extended the electromagnetic spectrum but, much more, because it enabled us to see so much that we had been unable to see. The word "fundamental" in basic science, which is often used as a synonym for "important," can be partly paraphrased as "relevance to neighboring areas of science." I would therefore sharpen the criterion of scientific merit by proposing that, other things being equal, *that field has the most scientific merit which contributes most heavily to and illuminates most brightly its neighboring scientific disciplines*. This is the justification for my suggestion to make it socially acceptable for people in *related* fields to offer opinions on the scientific merit of work in a given field. In a sense, what I am trying to do is to extend to basic research a practice that is customary in applied science. A project director attempting to get a reactor built on schedule is expected to judge the usefulness of component development and fundamental research which bear on his problems. He is not always right, but his opinions usually are helpful both to the researcher and to the management disbursing funds.

I turn now to the most controversial criterion of all: social merit or relevance to human welfare and the values of man. Two difficulties face us when we try to clarify the criterion of social merit. First, who is to define the values of man, or even the values of our society? Second, just as we shall have difficulty deciding whether a proposed research helps other branches of science or technology, so we shall have even greater trouble deciding whether a given scientific or technical enterprise indeed furthers our pursuit of social ends, even when those ends have been identified. With some ends we have little trouble: adequate defense, or more food, or less sickness, for example, are rather uncontroversial. Moreover, since such goals are relatively easy to describe, we can often guess whether a scientific activity is likely to be relevant, if not actually helpful, in achieving one of them. On the other hand, some social values are much harder to define; perhaps the most difficult is national prestige. How do we measure national prestige? What is meant when we say that a man on the moon enhances our national prestige? Does it enhance our prestige more than, say, discovering a polio vaccine or winning more Nobel Prizes than any other country? Whether or not a given achievement confers prestige probably depends as much on the publicity that accompanies the achievement as it does on its intrinsic value.

Among the most attractive social values that science can help to achieve is international understanding and cooperation. It is a commonplace that the standards and loyalties of science are transnational. A new element has recently been injected by the advent of scientific research of such costliness that now it is prudent as well as efficient to participate in some form of international cooperation. The very big accelerators are so expensive that international laboratories such as CERN at Geneva are set up to enable several countries to share costs that are too heavy for them to bear separately. Even if we were not committed to improving international relations, we would be impelled to cooperate merely to save money.

Bigness is an advantage rather than a disadvantage if science is to be used as an instrument of international cooperation; a $500 million cooperative scientific venture, such as the proposed 1000 GeV intercontinental accelerator, is likely to have more impact than a $500,000 Van de Graaff machine. The most expensive of all scientific or quasi-scientific enterprises, the exploration of space, is from this viewpoint the best-suited instrument for international cooperation. One can only hope that the cooperation between the United States and the U.S.S.R. in exploring space, tentative and halting as it is, will be greatly expanded and strengthened.

Some Specific Fields Assessed

Having set forth these criteria and recognizing that judgments are fraught with difficulty, I propose to assess five different scientific and technical fields: molecular biology, high-energy physics, nuclear energy, manned space exploration, and the behavioral sciences. Two of these fields, molecular biology and high-energy physics, are generally regarded to be basic sciences; nuclear energy is applied science; the behavioral sciences are a mixture of both applied and basic science. Manned exploration of space, though it requires the tools of science and is popularly regarded as being part of science, has not yet been proved to be more than quasi-scientific, at best. The fields that I choose are incommensurable: how can one measure the merit of behavioral sciences and nuclear energy by the same scale of values? Yet the choices between scientific fields will eventually have to be made whether we like it or not. Criteria for scientific choice will be most useful only if they can be applied to seemingly incommensurable situations. The validity of my proposed criteria depends on how well these criteria can serve in comparing fields that are hard to compare.

Of the scientific fields now receiving public support, perhaps the most successful is molecular biology. Hardly a month goes by without a stunning success in molecular biology being reported in the *Proceedings of the National Academy of Sciences*. A most striking recent advance has been the cracking by M. Nirenberg and S. Ochoa of the code according to which triples of bases determine specific amino acids in living proteins. Here is a field which rates the highest grades as to its ripeness for exploitation and competence of its workers. Its viewpoints unify, to an extraordinary degree, large stretches of other biological sciences — genetics, cytology, microbiology — and therefore according to my criteria must be graded A+ for scientific merit. It also must be given a very high grade in social merit, and probably in technological or medical merit — more than, say, taxonomy or topology. Molecular biology is the most fundamental of all the biological sciences. With understanding of the manner of transmission of genetic information ought to come the insights necessary for the solution of such problems as cancer, birth defects, and viral diseases. Altogether, molecular biology ought, in my opinion, to receive as much public support as can possibly be pumped into it. Since money is not limiting its growth, many more postgraduate students and research fellows in molecular biology ought to be subsidized so that the attack on this frontier can be expanded as rapidly as possible.

The second field to be considered is high-energy physics. This field of endeavor originally was motivated most strongly by our desire to understand the nuclear force. In this it has been only modestly successful; instead, it has opened an undreamed-of subnuclear world of strange particles, a world in which mirror images are often reversed. The field has no end of interesting things; it knows how to do them, and its people are the best. Yet I would be bold enough to argue that, at least by the criteria which I have set forth — relevance to the science in which it is embedded, relevance to human affairs, and relevance to technology — high-energy physics rates poorly. The world of subnuclear particles seems to be remote from the rest of the physical sciences. Aside from the brilliant resolution of the τ-particle paradox, which led to the overthrow of the conservation of parity, and the studies of mesic atoms (which are not done at *ultra*-high energy), I know of few discoveries in ultra-high-energy physics which bear strongly on the rest of science. This view must be tempered by recognition of the fairly considerable indirect fallout from high-energy physics — for example, the use of strong focusing, the development of ultra-fast electronics, and the possibility of using machines like the Argonne ZGS as very strong, pulsed sources of neutrons for study of neutron cross sections.[8] As for its direct or immediate bearing

[8] I should interject at this point that, since this essay first appeared in *Minerva* in 1963, this particular point has provoked a great deal of criticism, particularly from the high-energy physicists who deny my claim that high-energy physics is "remote" from other branches of science. As V. F. Weisskopf says, "Questions concerning the stability of nucleons, the reasons for the mass difference of neutron and proton on which the existence of matter is based . . . the problem of why there is one and only one electric charge unit, are these and similar questions to be considered unimportant and remote from the rest of science? It seems that they aim at the center of all scientific interest." "Two Open Letters: Weisskopf/Weinberg," *Physics Today 17,* 46 (June 1964).

My response to this criticism is that, apart from cosmology (to which high-energy physics almost certainly will contribute), there are few branches of science that seem to be waiting breathlessly for results from high-energy physics to get on with their jobs. Most discoveries in high-energy physics, intrinsically exciting and interesting as they may be, will probably make little difference as far as *what is done* to elucidate the rest of the physical universe is concerned. If one compares high-energy physics with quantum mechanics, one senses a profound difference in this respect. The latter, being concerned in a relatively workaday fashion with familiar phenomena, served as a key to the understanding of vast parts of modern science. The former, being concerned with a much narrower segment of the physical universe, can hardly be expected to have so broad an impact on our scientific endeavor as did quantum mechanics. I therefore, admittedly in retrospect, see an *urgency* in the pursuit of quantum mechanics which seems to me lacking in the pursuit of high-energy physics. The "remoteness" of high-energy physics is analyzed by Gerald Feinberg in "Physics and the Thales Problem," *The Journal of Philosophy LXIII,* 5–17 (January 6, 1966).

on human welfare and on technology, I believe it is essentially nil. If high-energy physics were cheap, these two low grades would not bother me since I sense the intrinsic beauty and majesty of the entirely new phenomena uncovered by the high-energy physicist. But high-energy physics is expensive, not so much in money as in highly qualified people, especially those brilliant talents who could contribute so ably to other fields which contribute much more to the rest of science and to humanity than does high-energy physics. On the other hand, if high-energy physics could be strengthened as a vehicle for international cooperation, if the much-discussed intercontinental 1000 GeV accelerator could indeed be built as a joint enterprise between East and West, the expense of high-energy physics would become a virtue, and the enterprise would receive a higher grade in social merit than I would now be willing to assign to it.

The third field is nuclear energy, for which I have passion and aspiration, and therefore I am not unbiased. Being largely an applied effort, nuclear energy is very relevant to human welfare. As I have pointed out in Part I, we now realize that in the residual uranium and thorium of the earth's crust, mankind has an unlimited store of energy, enough to last for millions of years. With an effort of only one tenth of our manned space effort we could, within ten or fifteen years, develop the reactors that would tap this resource. Only rarely do we see ways of *permanently* satisfying one of man's major needs, in this case, energy. In high-conversion-ratio nuclear reactors, we have such means. Moreover, again as I explained in Part I, we begin to see ways of applying large reactors of this type to realize another great end, the economic desalting of the ocean. Thus the time is ripe for exploitation. Nuclear energy rates so highly in the categories of technical and social merit and timeliness that I believe it deserves strong support, even if it got low marks in the other two categories — its personnel and its relationship to the rest of science. Suffice it to say that in my opinion the scientific workers in the field of nuclear energy are good and that nuclear energy in its basic aspects has many ramifications in other scientific fields.

Next on the list are the behavioral sciences: psychology, sociology, political science, anthropology, and economics. The workers are of high capability; the sciences are significantly related to each other; they are deeply germane to every aspect of human existence. In these respects the sciences deserve very strong public support. On the other hand, it is not clear to me that the behavioral scientists, on the whole, see clearly how to attack the important problems of their sciences. Fortunately, the total sum involved in behavioral science research is now relatively small, as it well must be when what are lacking are deeply fruitful, and generally accepted, points of departure. I would hope that such points of departure will crop up and that we can soon devote much more to the social sciences than now seems appropriate to me.

Finally, I come to manned space exploration. The personnel in the program are competent and dedicated. With respect to ripeness for exploitation, the situation now seems fairly clear. Our "hardware" is in good shape, and we can expect it to get better — bigger and more reliable boosters, better communication systems, and so on. Even the human being's tolerance to the space environment (which I questioned when I first wrote this essay in 1962) seems to be much better than some of us had feared.

The main objection to spending so much manpower, as well as money, on space exploration is its remoteness from human affairs, not to say the rest of science. In this respect, space (the exploration of very large distances) and high-energy physics (the exploration of very small distances) are similar, though high-energy physics has the advantage of greater scientific validity. There are some who argue that the great adventure of man into space is not to be judged as science, but rather as a quasi-scientific enterprise, justified on the same grounds as those on which we justify other non-scientific national efforts. The weakness of this argument is that space requires many, many scientists and engineers, and these are badly needed for such matters as clarifying our civilian defense posture or, for that matter, working out the technical details of arms control and foreign aid. If space is ruled to be nonscientific, then it must be balanced against other nonscientific expenditures like highways, schools, or civil defense. If we do space research because of prestige, then we should ask whether we get more prestige from a man on the moon than from successful control of the water-logging problem in Pakistan's Indus Valley Basin. If we do space research because of its military implications, we ought to say so. Perhaps the military justification, at least for developing big boosters, is plausible, as the Soviet experience with rockets makes clear.

The Big Problem of "Big Science"

The main weight of my argument is that the most valid criteria for assessing scientific fields come from without rather than from within the scientific discipline that is being rated. This does not mean that only those scientific fields that have high technical merit or high social merit deserve priority. Scientific merit is as important as the other two criteria, but, as I have argued, scientific merit must be judged from the vantage point of the scientific fields in which each field is embedded rather from the vantage point of the field itself. If we support science to maximize our knowledge of the world around us, then we must give the highest priority to those scientific endeavors that have the most bearing on the rest of science.

The rather extreme view which I have taken presents difficulties in practice. The main trouble is that the bearing that one science has on another science so often is not appreciated until long after the original discoveries have been made. At the time Purcell and Bloch first discovered nuclear magnetic resonance, who was wise enough to guess that the method would become an important tool for unraveling chemical structures? Or how could one have guessed that Hahn and Strassmann's radiochemical studies would lead to nuclear energy? And, indeed, my colleagues in high-energy physics predict that what we learn about the world of strange particles will in an as yet undiscernible way teach us much about the rest of physics, not merely much about strange particles. They beg only for time to prove their point.

To this argument I say first that choices are always hard. It would be far simpler if the problem of scientific choice could be ignored, and possibly in some future millennium it can be. But there is also a more constructive response. The necessity for scientific choice arises in Big Science, not in Little Science. Just as our society supports artists and musicians on a small scale, so I strongly favor society's supporting science that rates zero on all the external criteria, provided it rates well on the internal criteria (ripeness and competence), and provided it is carried out on a relatively small scale. It is only when science makes serious demands on the resources of our society — when it becomes Big Science — that the question of choice really arises.

At present, with our society faced with so much unfinished and very pressing business, science can hardly be considered its major business. For scientists as a class to imply that science can, at this stage in human development, be made the main business of humanity is irresponsible — and, from the scientist's point of view, highly dangerous. It is quite conceivable that our society will tire of devoting so much of its wealth to science, especially if the implied promises held out when big projects are launched do not materialize in anything very useful. I shudder to think what would happen to science in general if our manned space venture turned out to be a failure, if it turned out, for example, that the brilliant docking maneuver demonstrated in 1965 proved to be much more difficult when done close to the moon. It is as much out of a prudent concern for their own survival, as for any loftier motive, that scientists must acquire the habit of scrutinizing what they do from a broader point of view than has been their custom. To do less could cause a popular reaction which would greatly damage mankind's most remarkable intellectual attainment — modern science — and the scientists who created it and must carry it forward.

2. *Criteria for Scientific Choice II: The Two Cultures*[9]

The Financial Support of Science as a Whole

In the previous essay I proposed criteria which could be invoked in judging how to allocate support to different, competing branches of basic science. Such allocations seem to be necessary because what society is willing to spend on all of science is not enough to satisfy every worthy claim on the total funds available for science. I turn now to the broader question: what criteria can society use in deciding how much it can allocate to science as a whole rather than to competing activities such as education, social security, foreign aid, and the like?

That such a question can assume any urgency is in itself remarkable. To have suggested that the federal government of the United States would be spending about 2.5 per cent of the Gross National Product for research and development would have been unbelievable twenty-five years ago. Most of the new attitude toward government support of science and technology was prompted by war and fear of war. In the mind of the public, scientific strength has been equated with military strength. Support of science at first was only dimly distinguished from support of the military. But this attitude is changing, partly because the thermonuclear stalemate seems

[9]*Minerva III*, 3–14 (Autumn 1964).

to have reduced our fear of war (at least with our major thermonuclear adversary, the U.S.S.R.), and partly because the fantastic successes of modern science have begun to penetrate the awareness of the public. Science per se, as a valid human activity supported by the public, has acquired some standing, possibly analogous to that of religion in the era before the separation of church and state. As science has become big, it has acquired imperatives, just like any other activity of government, to expand and to demand an increasing share of public resources, and now, for the first time, it has become big enough to compete seriously for money with other major activities of government.

The criteria for choice between different fields of basic science I proposed earlier were of two kinds, internal and external. Internal criteria could be established entirely within the scientific field being considered. External criteria could be established only from outside. My main point was that a good rating according to the internal criteria was a necessary but not sufficient condition for *large-scale* public support of a field of science. Only if a field rated highly according to criteria generated outside its own universe could it properly expect large-scale support by society.

Insofar as the support of science as a whole can be viewed as different from support of each of the separate branches and kinds of science, I believe one can apply analogous criteria. Society, in its support of science, assumes that science is a competent, responsible undertaking. But society is justified in asking more than this of science as a whole. However vaguely stated, society expects science somehow to serve certain social goals outside science itself. It applies criteria from without science — broadly, criteria concerned with human values — when it assesses the proper role of science as a whole relative to other activities. We scientists concede this implicitly when we agree that responsibility for choosing between science and other activities belongs primarily to the nonscientists, the members of legislative bodies or the head of the executive branch of government and his staff. In the language of Stephen Toulmin,[10] the choice between "science done for its own sake" and other activities of the society is a political choice, as contrasted to an administrative choice, and it is to be made by politicians.

The ordering of human values upon which such choices must ultimately be based is a philosophical question that I shall not discuss here. I shall assume that we have decided on social goals and shall then ask how we can translate these into practical recipes for deciding how much science we can afford.

The Budgetary Separation of Pure and Applied Science

I shall dispose of the question of what fraction of society's over-all effort should go into science as a whole by arguing, along with many others, that it is misleading to budget for "science as a whole." The basis for the claim which applied science makes on society is so different from that of pure science that lumping them together tends to cloud some of the issues.

Applied science is done to achieve certain ends which usually lie outside of science. In deciding how much we should allocate to a project in applied science, we at least implicitly assess whether we can achieve the particular end more effectively by scientific research than by some other means. For example, suppose we wish to control the growth of population in India, and suppose we have at our disposal $200 million per year for this purpose. We could devote most of this sum to investigating fertility, to developing better contraceptive techniques, or to studying relevant social structures in some Indian village. Or, alternatively, we could use the money to buy and distribute existing contraceptive equipment, such as the ring or the pill, perhaps using some of the

[10]"The Complexity of Scientific Choice: A Stocktaking," *Minerva II*, 343–359 (Spring 1964).

money as incentive payment to induce women to accept the technique. Which way we spend our money is a matter of tactics; evidently no general proposition can tell us how much of our effort ought to be spent on research rather than on practice in trying to achieve effective birth control in India. The scientific work that goes toward solving this problem ought to compete for money with alternative, nonscientific means of controlling the growth of population in India rather than with the study of, say, the genetic code. More generally, where a piece of research is done to further an end which society has identified as desirable, support for this type of scientific work should be considered as part of the bill for achieving the end, not as part of the "science budget." Only that scientific research which is pursued to further an end arising or lying within science itself should be included in our science budget.

This view has become quite popular in many recent discussions of the subject.[11] It is appealing to the scientist because setting support for applied science outside the science budget reduces the latter enormously, from $16 billion to perhaps $1.5 billion. At this level the whole question of choice between scientific and nonscientific activities becomes much less significant.

But this stratagem is not as clearly justified as it at first appears. Ruling applied science to be part of the budget of nonscientific activities, not of the scientific budget, does not eliminate competition between applied science and basic science. Applied science requires in supporting roles, by and large, the same kind of people as does basic science. Building a large accelerator engages electrical engineers who would otherwise be available to help design control systems for rockets. In allocating support for a given applied science, one must keep in mind the effect of such allocation on basic

[11]A particularly cogent presentation of this position is made by Stephen Toulmin, *op. cit.*

science, and in supporting basic science, one must keep in mind the effect on applied science. Edward Teller has argued that because of the great emphasis on basic sciences in our universities, we have created an atmosphere that is uncongenial to applied science. He insists that our important applied scientific undertakings suffer because we tend to direct our best talents to basic science, our not-quite-best to applied science. Though Teller's contention is difficult to prove, especially since we have had so many brilliant successes in applied science, my own experience supports his view: that we have been successful in our applied work despite a disconcerting bias against application.

A second difficulty is that the aim of any given branch of applied science tends to become diffuse as time goes on. The scientific work of any of the large mission-oriented government agencies started out specifically to further the mission of the agency. But as time has passed, these clearly defined, mission-oriented goals of applied scientific work have become less clear. Byways that originally were germane to the mission flourish even after they cease to be relevant. An investigation that began as a promising approach to solution of an applied problem ten years later becomes an interesting study pursued for its own sake, yet it continues to be described as applied science. Thus to omit applied science from the science budget would leave out a large amount of research which was at one time motivated by an extrascientific or applied end, but which is now pursued primarily because it is scientifically interesting to those carrying on the research.

Finally, the motivation for basic science is itself often less than pure. Is nuclear structure physics done to further science or to help build reactors? Is the structure of natural products pursued as a challenge to scientific virtuosity in organic chemistry or because out of such studies will come the knowledge of enzyme action which ultimately will lead to control of metabolic disorders? Thus consideration of support of basic research completely apart from applied research is not as clearly defined a proposition as many proponents of this position hold.

Nevertheless, I believe the *general* principle of considering the budget for applied research as outside our national science budget, and including only basic research in it, has one overriding advantage. By allocating funds to any applied research as a certain fraction of the budget of the (usually nonscientific) activity to which the research is intended to contribute, we keep straight our reasons for supporting the applied research. What this fraction for applied research should be must depend largely on internal criteria such as, Do we see ways of making progress, or, Are good research workers available? It probably should also depend on the impact that support of that field of applied research will have on neighboring basic fields.

The fraction of effort that goes into achievement of a broad end, such as aid to underdeveloped countries or national defense, by scientific research, instead of by engineering or by social action, can hardly be decided entirely by the scientists. The scientific-research approach to solutions of difficult social problems is becoming increasingly popular. Yet in at least some proposals for action of which I am aware, notably in foreign aid and control of world population, it seems that excessive claims were made for research, particularly research in the social sciences. Scientists, when asked to judge how to solve a complex social problem, more often than not recommend more science, just as high-energy physicists, when asked to recommend a program in basic science, will ask for more high-energy physics, or oceanographers for more oceanography. To overstate the capacities of scientific research as a technique for settling difficult social questions is no more sensible than it is to understate them.[12] Thus, just

[12]The possibility that new technologies may help resolve difficult social problems must of course be recognized; I return to this point in Part IV.

as I have argued that scientific panels, judging how much money should be allocated to one branch of science rather than to another, should include representatives of neighboring branches of science, so a panel determining how much scientific research rather than engineering or production will best achieve a certain nonscientific end should include nonscientists as well as scientists.

Support for Basic Science as a Branch of High Culture

I have argued in the foregoing that applied and basic science should have separate budgets and that the budget for applied science should be set as a certain fraction of the effort allocated to the end (usually nonscientific) which applied science furthers. To this extent I have avoided the problem of choice between science as a whole and other human activities by denying the usefulness of the concept science as a whole. This still leaves unresolved the question of basic science, the science which cannot be justified by any reason except that it satisfies human curiosity. Are there some broad social ends, outside of basic science, which basic science serves, and to which its budget can be tied?

Obviously, some parts of basic science are important to applied science; in my view a much larger fraction of basic science is germane to applied science than many of my basic scientific colleagues are willing to concede. Most of the biological sciences are, in a sense, applied. For example, the most recondite and ingenious elucidation of the genetic map of *E. coli* is germane to the whole question of genetic abnormalities. Or again, plasma physics, a purely basic science, is central to thermonuclear research, an applied science which is pursued because we wish to enlarge mankind's energy resources.

It is natural to propose that such basic research receive a certain fraction of the resources going into the applied research which the basic research underlies. Every good applied research laboratory allocates to basic research a certain fraction of the resources granted to it for its related applied research. The ratio of basic to applied research frequently is very high and is often particularly high in those applied laboratories, such as the Bell Laboratories, which have had the most success in accomplishing their technological missions. I suggest that on a national scale, also, basic research be considered as a fixed charge or overhead on the applied research effort, wherever the basic research is intended to contribute to a field of applied science. In making an assessment of relevance, I would incline toward a broad interpretation; for example, I would consider most research in biology as a proper overhead charge to be assessed against the resources allocated to agricultural and medical research.

But what about those fields of basic research, a few of them very expensive, which are really very remote from any applied scientific problems, which are pursued primarily because the researchers find the science intensely interesting, and often because the findings in this field are likely to illuminate neighboring branches of basic science? To what can we tie the allocation of effort for such activities?

This is the most difficult of all the questions concerning public support of science, and any proposed solution must be put forward most tentatively. For basic science of this kind is primarily a somewhat disengaged intellectual activity, in the same sense as are music, literature, and art. Indeed, the analogy between the creative arts and this purest kind of basic science is sufficiently great to suggest that, insofar as it must make the choice, society might choose between the pure basic sciences on the one hand and the creative arts on the other. In allocating support for the purest basic research, our allocations for the other creative activities of man might be taken as a guide.

There are many analogies between the purest basic research activity and artistic activity. Each is an intensely individual experience, the effect of which transcends itself. The product of each is immortal — the theory of relativity, just as surely as *Hamlet* or the *Mona Lisa*. Each is concerned with truth — the highest of human manifestations; the one with scientific truth (which deals with the regularities in human experience), the other with artistic truth (which deals with the individuality of human experience).[13] Each enriches our life in unmeasurable though highly significant ways. Each belongs not only to its creator or discoverer, but to all mankind.

In a competition for support between pure science and the arts, I see two major arguments, one that supports the claim of science and the other the claim of art. The argument that favors science (aside from the obvious one, to which I shall return, that even the remotest pure science may eventually have practical application) is that scientific truth, being based on what we observe in nature, is publicly verifiable, whereas artistic truth, not subject to the same kind of control, is not publicly verifiable. Artistic critics disagree just as often as they agree. They have no objective and impartial arbiter, nature, to say what is true and what is not true. The truth of science, on the other hand, is rigorously and publicly tested by experiment or by observation or, in the case of mathematics, by logic. As I have already suggested, scientific criticism weeds out scientific nonsense more efficiently than artistic criticism weeds out artistic nonsense. Scientific research and thought, in their mutual and ruthless criticism that aims ever more strongly toward a whole and consistent structure, are embedded in what Michael Polanyi has called the "Republic of Science":[14] the entire scientific community, whose mutual interaction is governed by rules of conduct that are themselves laid down by nature, the great scientific

[13]This point was illuminated for me in Jacques Barzun's *Science: The Glorious Entertainment*, 227 *et seq.*, Harper & Row, New York (1964).

[14]M. Polanyi, "The Republic of Science: Its Political and Economic Theory," *Minerva I*, 54–73 (Autumn 1962).

lawgiver. The republic of science forces science to be a responsible undertaking, at least in the sense that what science does is true and, in some approximation, true forever. The corresponding republic of the arts has no such final arbiter that can force art to be as reponsible as science. Insofar as public support ought to go for the more responsible undertaking, the purest science in this regard merits more support than do the arts.

But there is another argument which at present favors the arts. Pure science — that is, science which does not have foreseeable practical applications, such as elementary particle physics or cosmology — is by and large an arcane enterprise which is appreciated mainly by its practitioners. The arts, on the other hand, generally have a larger audience. Many more people in the world today can enjoy Beethoven's Ninth Symphony than can enjoy Schroedinger's paper on quantization as an eigen-value problem. Granted that the intellectual delight experienced by the creator in pure science matches that of the creator in art, the direct products of the artist's efforts give more enjoyment to more people than do the products of the scientist. Of course, to the degree that even the purest science may eventually result in practical applications, it too affects the public; but we are speaking here of basic science without regard to possible practical application.

The well-paid pure scientists among my readers will undoubtedly object to being thus converted into scientific bohemians shivering in poorly heated garrets. But pure science is not doomed to that poor an existence if our society decides, even now, to support it on about the same scale as it supports the arts. It is true that the arts are supported poorly by government, but the total amount paid by society (that is, private individuals and associations, and local and federal governments) for the arts is not negligible, and the support is growing. In estimating the total support we give to the arts, we must include the value of theater admissions, the value of books, better magazines and records, the total that goes to our performing arts, as well as the direct subsidies in the form of grants to creative artists. The total spent by the United States on all activities that one way or another are concerned with the arts amounted in 1960 to around $2500 million.[15] Only a fraction of this amount is spent directly by the federal government, but this is not relevant. Pure science, unlike music or literature, produces no directly salable commodity, and so if it is to be supported at all by the public, it must be supported by the public through its government.

Moreover, it seems likely, with the increase in leisure and the decrease in the amount we spend on armaments, that a larger and larger fraction of our national income ought to go to the arts. And indeed, the establishment in 1964 of the National Humanities Foundation, paralleling the National Science Foundation, presages just this trend. As I suggested in Part I, to make of pure science an avenue for expression of our creative intellectual energy, quite parallel to August Heckscher's proposal to make of the arts such an instrument,[16] strikes me as highly appealing. This last viewpoint was stated eloquently by N. N. Semenov,[17] the Soviet chemist. He visualizes science in the world of the future being appreciated and practiced as widely as are the arts in the world today, with every man a scientist to the extent of his intellectual capacity.

I put forward the idea, as only one among several possibilities, that the purest science be supported in the same spirit and at roughly the same level as are the arts. The arts, after all, are not the only nonscientific activity which gives deep intellectual or spiritual satisfaction. For example, religion even today gives great spiritual satisfaction to many people —

[15]Estimate by A. Mitchell of Stanford Research Institute, as reported in *Business Week,* p. 68 (January 19, 1963).

[16]August Heckscher, "The Arts in the 1980's," an Edward Austin Sheldon Lecture presented as one of a series on "Possibilities of the Eighties," Oswego State University, Oswego, New York (1964).

[17]N. N. Semenov, "The World of the Future," *The Bulletin of the Atomic Scientists XX,* 10–15 (February 1964). The same idea was expressed by George Bernard Shaw in *Back to Methuselah.*

in our country, to many more than do the arts or sciences. And, indeed, a case can be made for using the level of support of religion instead of art as a yardstick for how much pure science our society ought to support.

And yet, despite the analogies between science and art, or between science and religion, the idea of relating the degree of support of one to the degree of support of the other is somehow forced and artificial and not really satisfactory. In the long run, how much our society is going to spend on basic science depends upon the extent to which nonscientists develop the intellectual power and taste to appreciate, if not to discover, science. The question then is, Is it really likely that society will develop so congenial an attitude toward science, say as congenial an attitude as it now displays toward the arts or religion, that it will support the basic scientist at the level he thinks he needs?

Most scientists believe that society will be missing something very important should it not develop such an attitude toward pure science. Every scientist knows that much of the satisfaction he derives from his scientific career comes not only from his own original discoveries, but also from the thrill he experiences when he understands, for the first time, someone else's great discovery.

As I have already implied,[18] one need not have a great intellect to appreciate a scientific discovery, at least enough to give one real satisfaction. I would guess that all those intelligent enough to take a university degree could learn enough to appreciate some branch of science, if not the most sophisticated parts, then at least the simpler parts. Nor is it necessary for all the public to understand all of basic science. Just as science itself has fragmented under the pressure of the information explosion, so I visualize that lay scientists would also form somewhat separate communities. Perhaps there might develop the equivalent of molecular biology fan

[18]Cf. Part I, page 36.

clubs, high-energy fan clubs, and oceanography fan clubs, even as we now have amateur astronomers, radio hams, and hi-fi enthusiasts.

To educate so many people to a point where they can achieve a sense of participation in the march of science poses a major problem. The scientists themselves will have to spend much effort conveying their message, in intelligible terms, to the rest of society. They will have to deal sympathetically (much more so than I think they now do) with the scientific popularizers and with the scientific educators. If the scientists and their parascientific associates are unable to convey this sense of scientific adventure to the community that supports them, I cannot see how the purest basic research can, in the long run, expect to receive the support it will demand in the future.

Support for Basic Science as an Overhead Charge on Applied Science and Technology

I confess to a residual skepticism about our society's acquiring this sophistication in the short run, which means, for the working scientist, the years until his retirement. It is probably utopian to expect every man in the street to become an amateur scientist or even a science fan.

Thus, much as I hope that our society will acquire this degree of scientific sophistication, it seems clear that in the near, as opposed to the distant, future we shall have to present a more realistic claim on society for support of basic science done for the sheer intellectual pleasure it affords its practitioners. I therefore return to my earlier suggestion that basic science in fields clearly relevant to applied science be viewed as an overhead charge on that particular applied science — that is, against the political mission the applied science is intended to accomplish. I would extend the idea and urge that the *purest* basic science be viewed as an overhead charge on the society's entire scientific and technical enterprise and that the charge be assessed on the whole activity because, in a general and indirect sort of way, such basic science is expected eventually to contribute to the whole technological system. In some cases the help will be direct, as when a discovery in cosmology illuminates a point in nuclear structure physics; in more cases the help will be indirect, as when a professor, whose research is in an abstruse field of mathematics, inspires a young engineering student with the beauties of the classical calculus of variations.

Some such view of the relation of the purest basic science to the entire technical enterprise was implicit in Executive Order 10521, issued in 1954 by President Eisenhower, concerning the terms of reference of the newly founded National Science Foundation:

> As now or hereafter authorized or permitted by law, the foundation shall be increasingly responsible for providing support by the Federal Government for general-purpose basic research through contracts and grants. The conduct and support by other federal agencies of basic research in areas which are closely related to their missions is recognized as important and desirable, especially in response to current national needs, and shall continue.

According to this view, the purest basic science, being an overhead assessed against the entire enterprise, would receive support at a level determined as a fraction of the entire enterprise. What this fraction should be would be a political decision; but if all such research were supported by a National Science Foundation, as suggested by the Executive Order of President Eisenhower, this political decision would amount each year to setting the budget of the National Science Foundation.[19] Of course this political decision would be

[19]This is the predominant view of the essayists who contributed to "Basic Research and National Goals," a Report to the Committee on Science and Astronautics, U.S. House of Representatives, by the National Academy of Sciences, U.S. Government Printing Office, Washington, D.C. (March 1965).

influenced in part by the public's attitude toward science; but it would also be influenced by the attitude of legislators, who are probably more inclined toward science than is the general public, since so much of the business of national legislative bodies now involves science and engineering in one way or another.

Where do the criteria of choice I previously proposed fit into such a scheme? As I see the matter, these criteria would be used both by mission-oriented agencies in making administrative decisions concerning different kinds of basic science, and by a body, such as the National Science Foundation, with very broad terms of reference, and independent of any technological aim, in choosing between different basic fields. Within each allocation of funds made for a politically defined task there will always be more claimants than there are funds, and choices will still have to be made. The beauty of the idea of basic research as a "scientific overhead" is that it reduces to a more manageable proportion the size of each allocation of funds for scientific research.

Thus I have turned a full circle. I began by asking how much "science as a whole" our society could afford. In developing my views, I have successively reduced the magnitude of science which competes with society's other activities; first, by viewing the costs of applied science to be overhead charges on the tasks it seeks to further, second, by viewing the costs of mission-related basic science to be an overhead charge on mission-related applied science, and now, by suggesting that the purest science be an overhead on the entire scientific-technological enterprise. This is not to say that I reject altogether the view of science as culture, a view which places science per se directly in competition with other cultural activities of the society. It is merely that, in the short term, basic science viewed as an overhead charge on technology is a more practical way of justifying basic science than is basic science viewed as an analogue of art. Until and unless our society acquires the sophistication needed to appreciate basic science adequately, we can hardly expect to find in the admittedly lofty view of science as culture a basis for support at the level which we scientists believe to be proper and in the best interests both of society and of the scientists.

3. *The Coming Age of Biomedical Science*[20]

The Urgency of Biomedical Research

In this essay I shall expand on a conclusion I reached in "Criteria for Scientific Choice": that of all the sciences now supported by our society, biomedical science ought to stand first. My purpose therefore is to illustrate how the criteria developed in that essay can be applied in a specific, concrete way.

We are, or ought to be, entering an age of biomedical science and biomedical technology that could rival in magnitude and richness the present age of physical science and physical technology. Whether we shall indeed enter this age will depend upon the attitude toward Big Biology adopted by biomedical scientists and government agencies that support biology. Whether the age of Big Biology will be truly rewarding will depend on the common sense and integrity of all who participate in this adventure.

Of all the bases for claiming large-scale public support for a scientific activity, the possibility of alleviating human disease through such activity is obviously one of the most compelling. Of all the sciences, the biomedical sciences are the only ones specifically aimed at, and relevant to, alleviation

[20]Most of this essay appeared in *Minerva IV*, 3–14 (Autumn 1965).

of man's elementary sufferings, disease and premature death. There is urgency of the most excruciating kind in getting on with this job. The assault on human disease, insofar as it may result in alleviation of immediate everyday human suffering, has an urgency comparable to the urgency with which a nation prosecutes a war. Indeed, I would draw an analogy in this regard between wartime research in physics and present-day research in the biomedical sciences.

This claim to urgency hardly can be matched by any of the other great fields of natural science. Certainly those fields that base their claims to support primarily on the promise of enlarging the human spirit have, to my mind, a less valid case for *urgency* than do those fields that base their claim on the possibility of curing or preventing human disease. SU(n) symmetry is magnificent and soul-satisfying to those who understand it; a cure for leukemia is more immediate in its benefit to mankind.

Are the biomedical sciences that relevant to the conquest of disease? To an applied scientist like me, this question seems absurd. What strikes an observer most about modern biology is how the new viewpoints have unified the subject. The genetic code appears to be universal. The dogma of protein synthesis — DNA, messenger RNA, transfer RNA, protein — seems to be valid in almost every life form. The same twenty-odd amino acids build proteins in bacteria, in mice, and in men. This unity suggests that most of what we learn about biological mechanisms in almost any animal is likely to have ultimate medical applications, whereas the same degree of relevance to application cannot be claimed for large parts of modern physics, or astronomy, or mathematics.

Derek de Solla Price has stressed that applied science tends to be older science, the science that excited the basic scientist of a previous generation.[21] Though I agree with his general

[21]"Is Technology Historically Independent of Science? A Study in

102

thesis that applied science is older science, I think the characteristic time lag varies from science to science. The "basic" biological sciences are so strongly connected with medicine, in my view, that I would *a priori* expect findings in basic biological science to move into experimental medicine with great rapidity, so rapidly that the distinction between applied and basic is very blurred. For example, the work on suppression of immunological mechanisms by radiation and the limited use of this technique in organ transplantation came almost simultaneously. The bulk of the basic work on the enhancement of radio-sensitivity caused by oxygen and the application of this technique to hyperbaric x-ray therapy were separated by hardly five years. Because in the biomedical sciences the distinction between pure and applied is rather irrelevant, as a matter of tactics I have argued that all of biomedical science be viewed as applied science. This hardly endears me to some of my good friends who consider themselves to be basic biological scientists. Yet from the point of view I am discussing here — the validity of the biomedical sciences' claim to urgency and therefore the validity of their claim to large-scale support from society — the position of biology is far stronger if it regards itself as fighting the war against disease rather than the war to enlarge the human spirit, worthy as is the latter.

If the biomedical sciences are viewed as applied sciences, aimed at alleviating disease, then in assessing their priority they should be judged not so much against other pure sciences but rather against alternative means of alleviating disease. The most obvious such alternative is medical practice, including treatment centers, hospitals, medical education, nursing care, and so on. I have already alluded to the competition between the demands of medical practice and the demands of medical research. I expect this competition, which presently favors research, to shift toward practice now that a study such

Statistical Historiography," *Technology and Culture 6,* 553–568 (1965).

as that by the President's Commission on Heart Disease, Cancer and Stroke[22] is bringing our country's attention to the need for more medical practice. My own view is that we need more biomedical science *and* more medical practice, and that the two, taken together, deserve very high priority in our society's allocation of resources.

The Ripeness of Biomedical Research

Relevant as is the aim of a science to achievement of a recognized human value — in the case of biology, to the elimination of human disease — this can only be a partial justification for large-scale public support. Before any scientific field can expect support on a very large scale it must be at a stage where large-scale public support is likely to produce useful results. Anyone who claims that biomedical science should become our number-one scientific priority must show that this field is likely to give fair return for support received.

In this respect the situation in the biomedical sciences at first sight seems to stand between certain of the physical sciences and the behavioral sciences. Judging by the criterion of direct relevance to human welfare, any ordering would almost surely place the behavioral sciences at least on a par with, if not above, the biomedical sciences; the more abstract physical sciences would almost surely rate below these. Judging by the criterion of ripeness for exploitation — that is, whether lack of large-scale support is mainly what is holding up progress — abstract physical science, like elementary particle physics or astronomy, is probably ahead of biomedical sciences, and behavioral sciences are much farther behind. This at least is the view one would gather from the strength of the plea for support made by the physical scien-

[22]The President's Commission on Heart Disease, Cancer and Stroke, "A National Program to Conquer Heart Disease, Cancer and Stroke," U.S. Government Printing Office, Washington, D.C. (December 1964).

tists, compared with the relative weakness of the plea we hear from the biomedical scientists. I think, however, that the biomedical scientists understate their case.

To begin with, as I have said, the war on human disease is a tangible war — more tangible, say, than our efforts to enlarge the human spirit — and it should be fought with the same attitudes we adopt when fighting a shooting war. We expect less return per dollar expended when fighting a war than when carrying on a less crucially important activity. So I would argue that, because of the importance of each victory in the battle against disease, we ought to be willing to get less per dollar spent on biomedical research than we are willing to get from expenditures on the more remote fields of science. We should stop putting more resources into the enterprise only when we have reached a stage of negative return — when more resources *reduce* the total useful output, not merely raise the unit cost of an increased total output. My attitude in this respect resembles that of one of our country's larger chemical companies. The president of this company once told me that where something very important like the design of a new plant was at stake, the company did not hesitate to put more than enough scientists and engineers on the job. Even though this led to duplication of effort, not to speak of a sense of frustration on the part of some of the men, this technical redundancy has paid off very well indeed.

I believe the biomedical sciences are not near the stage where additional large-scale support will *reduce* the over-all output of the entire enterprise. It is apparent even to the most casual observer that we are beginning to understand many of the life processes which for so long had been mysteries: the revolution in molecular biology; or the beautiful elucidation of the mechanism of nerve action; or the new insights into the genetic control of immune mechanisms; or the implication of viruses in some cancers, notably animal leukemia, although the role of viruses had long been suspected. One can hardly believe that the many fruitful points of departure uncovered during the past decade are anywhere close to being exploited, or that, if more well-trained, well-supported investigators are set to work, new and startling points of departure will not be found.

Moreover, I think the biomedical sciences can be force-fed, even more than they are now being force-fed. More money for biology has raised the salaries of biologists, at least in the United States, so that presently biologists enjoy an unaccustomed affluence. Though this state of affairs annoys administrators, particularly of multidisciplinary laboratories where disciplines use each other's salary schedules as ratchets, the over-all effect as far as biomedical science is concerned is, on balance, good. More bright youngsters are attracted to well-paid careers than to poorly paid ones. In the United States such force-feeding of a discipline in the past has produced results. For example, the Atomic Energy Commission, by pouring money into nuclear research, caused it to flourish, and encouraged many young science students to go into nuclear research. Or again, the AEC and the U.S. Department of Defense deliberately established about a dozen interdisciplinary materials research laboratories; though it is too early to say positively, my impression is that materials research in the United States has profited by this action.

The New Style of Biomedical Research

There are other reasons, intrinsic to the changing style of research in biology, why more money will be needed. Most obvious is the growing cost of equipment. A modern electron microscope now costs $40,000, and more and more, cellular biology seems to depend on the electron microscope. Even now attempts are being made, both at Oak Ridge National Laboratory and at Argonne National Laboratory, to develop an electron microscope with a resolution of 1 Å. Such a device, if successful, would enable one to identify individual atoms in biological molecules. It might cost several million dollars.

But there are other, possibly subtler, reasons why biological research is becoming more expensive and is requiring more people. In earlier times, when biology was par excellence Little Science, biologists were content to look only at those problems that could be handled in the manner of Little Science. Genetics was done with fruit flies, with their large chromosomes, because fruit flies are inexpensive, not because fruit flies are as much like man as are mammals. Those questions that required large protocols of expensive animals were answered poorly or not at all, not because the questions were unimportant, but because to answer them was expensive and required the style of Big Science, which was so foreign to the biologists' tradition.

But this is changing, in part at least, because the Big Scientists from neighboring fields have taught the habit of Big Science to the biologists. Perhaps the best-known example of the drastically changed style of some biological research is the large-scale mouse genetics experiment of W. L. Russell at Oak Ridge. For the past sixteen years Russell has been studying the genetic effects of ionizing radiation in a mammal, the mouse. Since mutations even at high dose rates are so rare, Russell uses colonies containing 100,000 mice. To perform such experiments takes much money and many people, and yet it seems impossible to visualize any other way of obtaining the data.

The problem of large protocols that Russell faced, and the AEC solved (at a cost of a million dollars per year for this single experiment), is one which arises in many other situations. The increasingly important matter of low-level physical and chemical insults to the biosphere, or of subtle environmental factors in general, will require many large experiments if we are to assess accurately the various hazards that now bombard us. Or take old age, the commonest "disease" of all. Merely because the effects are subtle and often appear haphazardly, the study of aging requires large and expensive protocols. The tradition of the biologists has been to scrimp;

biomedical research avoided expensive experiments even if expensive experiments were required to obtain reliable statistics. Biology, while continuing its tradition of Little Science, will have to accept also the style of Big Science. Even though this is expensive, the biologists will find the public willing to support them.

There is another trend in the style of biology which will add to its expense. I refer to the increasingly interdisciplinary character of modern biology, and particularly its increasing dependence on the techniques and methods of the physical sciences and even of the engineering sciences. A few examples, taken from our experience at Oak Ridge, will illustrate these points. For instance, in attacking the problem of radiation insult, we have mobilized biochemists, cytologists, geneticists, pathologists, and biophysicists. Our dependence on disciplines even farther removed from biology is growing. Thus, our biochemists need large quantities of transfer-RNA, preferably separated into unique fractions, to study how amino acids are assembled into proteins. The problem in many ways is one in chemical engineering, and some of the chemical engineers at Oak Ridge have pitched in to help. What the chemical engineers already have done strikes me as being rather impressive. They have been able to extract as much as 600 grams of pure t-RNA from 300 kilograms of *E. coli* by fractionating crude nucleic acids and then separating the extract into specific transfer-RNA's by using a liquid ion-exchange system of the general sort developed in refining uranium ores. The resulting separations are superior to any that have been achieved by older methods.

Second, I mention, again from our Oak Ridge experience, the exciting developments in zonal centrifugation applied to biology. For many years very high-speed, very large, continuously fed centrifuges have been developed for separating the isotopes of uranium. Much of this work has been carried out at the K-25 Gaseous Diffusion Plant. Around 1959, N. G. Anderson of the ORNL Biology Division realized that such centrifuges, suitably modified, might separate cellular moieties on a larger scale than could be done with any other technique. And indeed, with the generous support of the National Cancer Institute and the AEC, this is exactly what has happened. With these centrifuges Anderson has been able to detect virus-like particles in leukemic blood more consistently than have most other investigators who do not have this tool. I would expect Anderson's centrifuges to become widely used in biomedical research, even though some of his centrifuges cost as much as $45,000.

I could list many other instances of the growing interaction between the biological sciences and the physical and engineering sciences — for example, the technique of medical scintillation spectrometry, which has become a medical speciality in its own right, or the wide use of computers in biomedical science, or, for that matter, the application of the methods of quantum chemistry to the attempts to understand the carcinogenic action of aromatic hydrocarbons. But I believe I have given enough examples to bring out the main points: that biomedical science is becoming ever more interdisciplinary, that the disciplines and techniques it draws upon are expensive, and that this will add to the expense of biomedical science.

The New Biomedical Institutes

The changing style of biomedical research and its great and urgent expansion will affect the future organization of such research. At present, a very large part of biomedical research is carried out at universities, institutions that are, or should be, committed to education at least as strongly as they are committed to research. University biomedical research must flourish and must grow. We shall have to maintain Little Biology as well as Big Biology, and we shall have to produce many more trained biomedical scientists if we are to attack, with both styles, the problem of human disease.

But much of the great expansion in biomedical research should take place in biomedical research institutes, many of which will be directly affiliated with universities, but many of which will not. For, as Rossi[23] put it, and as I shall discuss in detail in Part IV, the social ecology of the university is not so well suited to a massive attack aimed at a single goal as is the ecology of the research institute. In the first place, the traditional departmental structure of the university is poorly suited to interdisciplinary approaches. In the second place, individuality and academic freedom in the university are preciously guarded prerogatives; these are often incompatible with achieving success in tasks that require cooperation.

The ecology of the research institute has a different tone: it is more hierarchical, its members interact with one another more strongly, and it is interdisciplinary. In the individualistic, competitive university environment, genius flourishes; but things go slowly because each genius works by himself with his own small group of students and assistants. In the less individualistic, cooperative institute environment, genius probably does not flourish as well, but things go very fast because so many different talents can be brought to bear on a given problem. It is a place in which, however, a single very able man can exert much more power and influence than he can in the university environment; it is a place where the whole is often much more than the sum of its parts.

If one accepts the proposition that biomedical science ought to be pursued with the same urgency with which we pursue military research, then the institute provides a better setting for such activity than does the university. In this I admit to being very much influenced by our own experience at Oak Ridge. There we have a prototype of a big biomedical institute; its central theme is the radiation insult to the bio-

[23]Peter H. Rossi, "Researchers, Scholars and Policy Makers: The Politics of Large Scale Research," *Daedalus 93*, 1142–1161 (Fall 1964). See also Alvin M. Weinberg, "But Is The Teacher Also a Citizen?", *Science 149*, 601–606 (August 6, 1965).

sphere. In the pursuit of this major theme, many disciplines are brought to bear. The enterprise is hierarchical (though benevolently so); it is big, it is interdisciplinary, and it is effective.

I would therefore suggest that much of the big expansion in biomedical research ought to go toward establishing additional interdisciplinary institutes like the Sloan-Kettering Institute or the contemplated environmental health institute of the World Health Organization. Certainly close connections with the universities are desirable; but I do not regard these as primary. The main job is to alleviate human suffering by learning as much as possible in as short a time as possible. In some cases this aim is furthered by close association with a university. I suspect that there are many cases where only a loose university affiliation is indicated.

The Indebtedness of Biology to the Physical Sciences

The coming Age of Biomedical Science will impose on administrators of biomedical research a new and unaccustomed responsibility to the physical sciences. I have already alluded to the increasing relevance of the physical sciences to the biomedical sciences. It is time, I suggest, for the community of biomedical science to recognize its dependence upon certain of the physical sciences and to assume a proper share of their support.

Support of certain parts of physical science has already been taken up by the biomedical sciences. For example, in the United States, the National Institutes of Health is now the largest single supporter of basic chemical research in the universities. But my impression is that such support tends to be somewhat bound by narrow interpretations of relevance.

I believe research in many of the physical sciences, such as structural organic chemistry, or x-ray and neutron diffraction, or even certain parts of solid state physics, is properly the concern of the biological sciences. The whole Watson-Crick development would have been impossible had it not been for major developments in the techniques of x-ray diffraction. Moreover, more and more of the world's leading biologists seem to be coming from the physical sciences. I mention, for example, Francis Crick, or Seymour Benzer, or Paul Doty, or Kenneth Cole. The debt owed to the physical sciences by the biomedical sciences is one of long standing, and it is growing. It is time for the biomedical sciences to begin repaying this debt.

The basic physical sciences in the United States are facing a major financial crisis. In the past they have been supported largely by three agencies: Department of Defense, Atomic Energy Commission, and National Aeronautics and Space Administration. But the missions of these agencies — defense, atomic energy, and exploration of space — are not likely to receive increasing support;[24] on the contrary, our country in the past year has made the political decision to keep these agencies at about their present level or even to reduce them somewhat. Thus the physical sciences, insofar as they are supported because they are relevant to the achievement of the missions of these agencies, are probably destined to receive relatively less support in the future than they have in the past.

But this predicament comes at the time when I would hope support for the biosciences will greatly increase, and when the connections between the physical and the biomedical sciences will become ever stronger. What is more natural than to ask the biomedical sciences to carry a fair share of the burden for supporting the many branches of physical science that are broadly relevant to the biomedical sciences? Such a plea from the hard-pressed physical scientists has justice on its side. I hope that the biomedical administrators, in their newly found affluence, will heed these cries from their colleagues in

[24]The situation in Vietnam has changed this as far as DOD is concerned. DOD's total budget has grown, but the war has reduced the money available to DOD for long-range research.

the physical sciences who have given them a free ride for so many years.

The Biologist as a Big Scientist

Traditional biologists must surely recoil in horror at the advice I give them: to expand even at the cost of individual effectiveness as long as their total output increases; to break down their traditional disciplinary barriers and to adopt more of the institute style of research as contrasted to that of the university; to overcome their suspicion of the physical scientists — in short, to accept the new style of Big Science in addition to the old style of Little Science.

But if horrified recoil is their reaction, I remind them that insofar as what they do is part of the war against human suffering, their desires and tastes are not all that matter. Biomedical science is not done, or, more important, is not supported by the public simply because it gives intense satisfaction to the dedicated and successful biomedical researcher. It is supported on a really large scale because out of it have come means of eliminating man's infirmities. If a style that runs counter to the traditional style is needed to mount a much larger biomedical research enterprise, then this style will have to be adopted, much as it hurts the sensibilities of those attached to the traditional patterns of scientific organization.

And, indeed, I have inveighed against the dangers of Big Science:[25] its too-frequent preoccupation with the big announcement rather than the big discovery, its tendency to substitute money for thought, its overabundance of administrators, its incompatibility with the educational process, even its inefficiency. As Sir Winston Churchill said, "I do not unsay one word of this." And nothing I have said implies that I

[25]"The Impact of Large-Scale Science on the United States," *Science 134*, 161–164 (July 21, 1961).

consider the style of Little Science to be obsolete. In urging more biomedical science, I plead both for more Big Science and for more Little Science.

Big Science, with all its dangers, does have a real place in the scheme of things. When the end to be achieved is important enough, and when the state of the science suggests that more support will lead to more results (and both these circumstances apply to biomedical science), then we are justified in going all out in our plea for public support. The coming Age of Biomedical Science will not be an unmixed blessing for the biologist; he surely will fret at being involved in something big and unwieldy and at times inefficient. Nevertheless, as a responsible member of the human race who is sensitive to the purpose of enlightened human activities such as biomedical research, he will have to submerge his instinctive distaste for bigness in the interest of humanity's welfare.

4. *Scientific Choice and Human Values*[26]

Often in human situations requiring a choice between alternatives, we are not confronted with all the alternatives simultaneously; we are therefore not called upon to order our choices very systematically. We decide on foreign aid first and farmers' aid later; we appropriate money for highways and money for hospitals rather independently. By contrast, our scientific alternatives present themselves to us simultaneously, and therefore oblige us to choose much more explicitly. The nuclear physicists, the astrophysicists, the oceanographers, the meteorologists, and the followers of many other scientific disciplines now confront our instruments of decision all at once, explicitly and urgently. Any philosophic underpinning that we may try to give as a guide in making such scientific choices runs the risk, unusual among philosophic doctrines, of actually being put to practical use.

In approaching the problem of scientific choice, I have tried in the previous essays to establish a calculus of scientific values, a scientific value system. I have subjected science to the same kind of criticism that traditionally is reserved for the arts or, to some extent, for every human activity whose support is subject to the give and take of politics. In the debate on scientific choice, we are asking of science, perhaps for the first time on a big scale, not Is this science *true* and does it

[26]Part of this essay appeared in the *Bulletin of the Atomic Scientists XXII*, 8–13 (April 1966).

add to man's knowledge? but rather, Is this science *worthwhile,* more worthwhile than another science? We are, in other words, confronted with the problem of clarifying and analyzing what is meant by scientific taste or scientific value.

The underlying assumption of my calculus of scientific values is that we cannot evaluate a universe of scientific discourse by criteria that arise solely from within that universe. Rather, we find that to make a value judgment, we must view the enterprise from a broader point of view than is afforded by the universe itself. This implies that I have assumed a new ethical principle for science: not only must science seek truth, it must seek *relatedness.* The value of chemistry as a field can hardly be decided by the chemists alone; they must ask the biologists who need the results of structural chemistry to elucidate the genetic mechanisms of the cell; or they must ask the physicists who cannot probe nuclear magnetic resonances unless they understand how the chemical environment affects the details of their NMR signals; or they must ask the reactor technologist who needs the chemistry of protactinium to design a continuous purification system in a thorium breeder. And so it is with the rest of science. The scientific merit of a field must be judged in large part by the contribution it makes, by the illumination it affords, and by the cohesion it produces in the neighboring fields.

To repeat, in asking of a basic scientific activity that it strengthen the unity of science, I am imposing a new demand on science. Not only must a branch of science seek the truth: to qualify as good science it must also help create unity from diversity; it must try to create deeper meanings which show relations between it and other branches of science. As Jacob Bronowski put it so beautifully in his *Science and Human Values,*[27] it is this creation of connections where none had previously existed, and not simply the unearthing of facts, that is the essence of scientific creativity. Bronowski had in

[27]Jacob Bronowski, *Science and Human Values* (Rev. Ed.), Harper & Row, New York (1965).

mind individual acts of scientific creativity; I have in mind the creation of new points of connection between a broad field of science and its neighbor.

What I plead for in imposing the criterion of relatedness on a scientific activity is a revolt against the fragmentation and specialization that bedevil our science and our society. The fragmented outlook has already permeated our universities, and in some instances it has begun to invade our elementary classrooms. I am realist enough to recognize that the trend is almost inevitable, but as the director of a large laboratory in which almost every scientific discipline is represented, and who gets increasingly frustrated with his inability to maintain a coherent picture of the whole, I shall continue to deplore the trend.

The Embeddedness of Human Values

The main philosophic principle that emerges from the very practical question, How should we cut the scientific pie?, is a reaffirmation of the embeddedness of scientific values, in the sense first that value judgments are to be made from a neighboring vantage point, and second, that such judgments are to depend on the extent to which the activity being judged enlarges or furthers the neighboring universes.

Can this calculus of scientific values and its underlying idea of embeddedness be applied in any useful or meaningful sense to human values generally? In short, can science, or more accurately, our attempt to construct an ethic for science, provide clues as to how to order human conduct or to create a system of human values? This is an old question which is usually answered in the negative; most writers argue that science, concerned with a nature which is morally neutral, is bereft of any capacity to make value judgments, to moralize if you will. But there have been contrary views, and I shall review two of them.

The first view, put forward by J. Bronowski, by A. Rapo-

117

port, and by M. Polanyi, argues that, because science and its practitioners are so strongly dedicated to truth, the entire enterprise of science — the "Republic of Science," as Polanyi calls it, or the "Society of Scientists," as Bronowski calls it — is the most perfect republic conceived by man. It is a social order as described by Rapoport "in which investigation, criticism, intellectual cross-fertilization, and intellectual revolution are always possible and always welcome."[28] In this social order truth reigns supreme. The style of scientific interaction — that is, the habit of self-criticism of the scientific community — keeps the quest for truth straight and narrow. The charlatan and the faker are ruthlessly exposed; the right collective path to follow emerges as the result of innumerable interplays between individuals, each playing the game according to rules that each understands and that keep everyone honest. Bronowski and Rapoport in effect propose that we devise a human ethic by copying the underlying ethic of the Republic of Science; in this sense, science provides a basis for human values and a system of human ethics.

I find this vision of an almost perfect Republic of Science, and of the correspondingly perfect Republic of Man drawn in its image, to be ever so appealing. Yet, aside from the practical question of actually making the supposedly isomorphic transition from the one republic to the other — drawing up the Constitution of the Republic of Man from the Constitution of the Republic of Science, so to speak — I see grave questions. These have mainly to do with the corruption of the Republic of Science caused by the advent of Big Science. In this age of big money for science, it is harder than in earlier days to find the scientist dedicated solely to truth; he is responsible for spending Big Money, and his pursuit of science is sometimes distorted by his method of funding. But more important is the blunting of science's instrument of self-criticism, the scientific literature, simply because science

[28]Anatol Rapoport, *Science and the Goals of Man*, 232, Harper and Brothers, New York (1950).

has become so big. The fragmentation of science tends to make scientific criticism parochial and even self-serving, a situation that is aggravated by the vastness of the inadequately read scientific literature. If the Republic of Science is to be kept honest by the traditional interplay of scientific critics, mediated by the scientific journals, what does the republic do when there are too many journals to read? Thus, appealing though the Republic of Science may be, it is hardly as perfect as Professor Polanyi or Professor Bronowski consider it to be; and it therefore can serve only as a partial basis for a human ethic as proposed by Bronowski and Rapoport.

The Imperative Toward Order

Another attempt to find within science a basis for a human ethic was put forward by Professor Bruce Lindsay of Brown University in 1959.[29] Lindsay points out that the living biological organism tends to reduce its local entropy: it tends to produce order out of disorder locally, although, of course, the total entropy of the universe continues to increase. As Lindsay views it, it is inherent in the biological organism, simply because it is a biological organism, to increase the order of what it incorporates. This in a way can be extended into an ethical principle: whatever increases order is natural in the sense of being biological. Ethics, insofar as it implies that that is good which increases order, is not so much a philosophical principle as it is a biological fact.

I find Lindsay's idea attractive, but the issue obviously revolves around what we mean by order in this context. A clue to a proper definition is suggested by a remark Eugene Wigner once made to me about his motivation in creating the majestic principles of symmetry which now dominate so much of physics: "I hope that I leave this world a little more orderly than I found it." By order here he means not a lock-

[29]R. B. Lindsay, "Entropy Consumption and Values in Physical Science," *The American Scientist 47,* 376 ff. (1959).

step Hitlerism devoid of human dignity. Rather he means that he hopes to increase the relatedness of things, to show, for example, how the existence of the Ω^- particle follows from a fairly natural extension of the symmetry principles which underlie much of atomic quantum mechanics.

Thus insofar as Lindsay's imperative toward order can be paraphrased into a quest for relatedness, for showing the connections between things and thus striving for a unity, I would be glad to accept his entropy principle as a basis for human ethical systems. But this seems to me to be a manifestation of the "principle of the embeddedness of values," which I believe underlies the proper analysis of our debate on scientific values. In deciding what science is good, we have invoked the notion that scientific merit is to be measured by the degree to which that science illuminates and deepens our understanding of the neighboring sciences. We have based our scientific value system, our scientific ethic so to speak, on the notion that truth is whole, that the purpose of science is not merely to unearth the facts but also to show the relatedness of facts. We have elevated the principle of scientific parsimony — that is, the idea that the aim of science is to unify our picture of the world and thus enable us to explain the world with the fewest ad hoc postulates — to an ethical or moral principle. Science, in striving for better relatedness between its segments, seeks to achieve more order, or, perhaps more accurately, Lindsay's imperative toward order is interpreted as an imperative toward relatedness.

There are some outside science who seem to regard relatedness as an ethical or aesthetic criterion. Sir Kenneth Clark, in his beautiful Smithsonian Bicentennial talk on artistic criticism,[30] said that "as science has become more specialized and more abstract, so art has felt the need to make its own realm purer and more inaccessible." He goes on to say "the notion

[30]"The Value of Art in an Expanding World," *Knowledge Among Men*, 55–57, Simon and Schuster, in cooperation with the Smithsonian Institution in Washington, New York (1966).

that this universally valid sense of form must be a sensation unconnected with other human experiences, was an error." In other words, Sir Kenneth seems to come to the same position with respect to artistic criticism as I have with respect to scientific criticism: some of the criteria of merit for judging art, no less than for judging science, must be sought outside science, and indeed, must be sought in the human life and spirit of which these two seemingly unrelated activities are manifestations.

And I would go further: I would argue that human value systems, not only in art and science but in the broadest human experiences — our sense of justice, our assessment of institutional purposes,[31] our criteria of virtue — involve the kind of ethical relativism implicit in our calculus of scientific values. Certainly everything that the social sciences have learned, particularly cultural anthropology, points to the idea that what is good can hardly be decided ultimately; what is good in one society may be bad in another and vice versa. That a similar principle underlies our judgment of scientific merit is not surprising. But our scientific calculus, combined with Lindsay's ethical imperative, allows us to move beyond a neutral and austere ethical relativism, that everything is a matter of taste, a matter of fashion. For we argue that things, scientific activities or human activities, are themselves embedded in a total human matrix, and that *merit is to be judged by the degree to which the activity, scientific or human, contributes to the unity and illumination, and ultimately to the harmony, of the many activities with which it interacts*. Thus, there is a kind of ethical reciprocity: we decide on the good from the standpoint of the neighboring universes; in making the judgment, we ask Does the activity or attitude we are judging help create a unity, a harmony in the universe doing the judging?

I recognize that my foray into the sticky realm of human

[31]I return to this point in Part IV.

ethics is tentative and halting. Yet I am encouraged by two separate points: first, that with the weakening of religion as a basis for modern ethical behavior, it would be well to have a substitute. We scientists are naturally attracted by attempts to find such substitutes in what is our professional business. And second, I am encouraged that similar views have been expressed by others. As Bronowski put it, "When Coleridge tried to define beauty, he turned always to one deep thought: beauty, he said, is 'unity in variety.' Science is nothing else than the search to discover unity in the wild variety of nature — or more exactly, in the variety of our experience."[32] Perhaps to use this same principle of unity in diversity as a meaningful ethical principle is asking too much. Yet, the analogy between our problem of scientific choice and the problem of human choice — that is to say, between the ethics of science and the ethics of man — is too tantalizing to be ignored. Perhaps such philosophic sallies will stimulate others to think about these issues and will encourage them to bring forth a logical ethic for science and a credible ethic for man.

[32]Jacob Bronowski, *op. cit.*, p. 16.

IV

THE INSTITUTIONS OF BIG SCIENCE

Today's federally supported scientific institutions are characteristically of two kinds. On the one hand there are the great federal laboratories, like the National Institutes of Health, or Oak Ridge National Laboratory, or the Space Technology Laboratories. These institutions derive all or most of their support from the federal government, though some are operated by private contractors, and some, like the Space Technology Laboratories, are even owned by private companies. These laboratories are mission-oriented, having been set up in the first place to help accomplish the missions of the government agencies that support them. These missions, like achieving adequate defense, or cheap energy, or better health, originate outside science; science is simply a means to help achieve nonscientific, politically defined aims. I shall lump all these institutions together as national laboratories, although this name is most properly applied to the atomic energy laboratories. In a few cases, where the mission of the sponsoring agency is itself understood to be the pursuit of a particular branch of science or technology, national laboratories have been set up to perform large and expensive basic research in pursuit of such scientific aims. The best example in our country of such an institution is the Brookhaven National Laboratory. Though a fair amount of applied work on reactors and radioisotope applications is done there, Brookhaven is pre-eminently a center for basic research in nuclear and elementary particle physics.

The other federally supported scientific institutions are what Clark Kerr calls the federal grant universities. These are large state or privately operated universities which depend upon the federal government for support of most of their research. The federal grant universities sometimes operate mission-oriented national laboratories for the government. For example, The University of Chicago operates the Argonne National Laboratory for the Atomic Energy Commission;[1] the Massachusetts Institute of Technology operates the Lincoln Laboratory for the U.S. Air Force. But, generally speaking, these large contractor establishments are only loosely connected with the universities. When we speak of government-sponsored research at a university we usually mean rather small-scale, individual research in pure science.

Characteristic of most university research is its discipline-orientation and purity, as contrasted with the mission-orientation and applicability of research at the national laboratories. In the university, ideally, problems are chosen by the research professor; in the national laboratory, problems are chosen by the research director. This of course oversimplifies the situation since a professor cannot always get support for what interests him, nor can a research director always easily persuade a balky scientist in his laboratory to do what he wants him to do. Moreover, the national laboratories do large amounts of basic research in support of their applied missions, and, in many cases, to exploit unique and expensive facilities that are too large or too hazardous to be placed on university campuses. But, by and large, this difference in aim, and therefore in tone and attitude, does prevail in the two kinds of scientific institutions.

The federal laboratory and the federal grant university

[1]Actually The University of Chicago and Associated Midwest Universities now share the responsibilities for operating the Argonne Laboratory.

complement each other. The discipline-orientation of the one complements the mission-orientation of the other; the university's primary concern with pure research complements the national laboratory's primary concern with applied research. Both kinds of institutions have enjoyed great success. American pure science is in most areas pre-eminent. Although our lead over the rest of the world is probably not as great in applied science, I believe we are ahead also in many of the major technologies. In atomic energy, computers and communications, space, and electronics, no nation has surpassed us.

Yet, in looking to the future, one must wonder about the continued, long-term prospect for these institutions. Both institutions, being supported by public funds, must somehow remain relevant to public purposes, however defined. The laboratories were set up to accomplish, by scientific means, certain politically defined missions; what happens to the laboratories when their missions are accomplished or by-passed? The universities are dedicated to increasing knowledge within the separate disciplines; what happens to the universities as the disciplines, responding to their own internal logic and force, become so remote from the rest of society that the public is no longer willing to support them? Thus for entirely different, rather complementary, reasons the two institutions must face the broad question, Can these publicly supported institutions display the flexibility needed to remain relevant to public purposes? If they cannot, it will be difficult for them to retain the public confidence and support they now enjoy.

1. *National Laboratories and Missions*[2]

The large, mission-oriented laboratories, like the National Bureau of Standards, or the atomic energy laboratories, or the Battelle Memorial Institute, are relatively new inventions. The oldest in our country, the National Bureau of Standards, was founded in 1901. World War II spawned the atomic energy laboratories, and caused the military laboratories to grow. Sputnik converted the old National Advisory Committee for Aeronautics Laboratories into the many times larger complex of space laboratories.

These laboratories are of two kinds: project and component laboratories. Project laboratories may be exemplified by the wartime Metallurgical Laboratory at Chicago, or the Marshall Space Flight Center at Huntsville. The Metallurgical Laboratory was responsible for an entire project: the controlled release of atomic energy, and the design of the Hanford chain reactors and the plutonium separation plants. The Marshall Space Flight Center has been responsible for specific pieces of space hardware, such as the Redstone missile and the Saturn rocket. Component laboratories, of which the original NACA laboratories are prime examples, investigate matters judged to be relevant to the successful

[2]Parts of this essay appeared under the title "Future Aims of Large Scale Research," *Chemical and Engineering News, 33,* 2188–2191 (May 23, 1955).

126

working of complete pieces of hardware, but do not themselves take responsibility for this hardware. Most large, mission-oriented laboratories work on both components and projects, but usually one or the other tends to dominate.

Many of these institutions have grown enormously since they were first established: Oak Ridge National Laboratory is four times larger than it was during World War II; the NASA Lewis Laboratory grew from 2722 in 1960 to 4697 in 1963; the population of the Argonne National Laboratory has grown from 1350 to 5300 since 1945. Yet, in spite of their growth, I believe the ultimate viability of these institutions depends upon, and is limited by, the existence of new problems of sufficient magnitude, scope, and interest to challenge them adequately.

It is an old story for scientists to think they discern limits to the problems that will prove interesting or useful to their successors. My purpose will be not so much to show the limits of existing problems as to point out the areas where new problems are likely to emerge: problems that will require mobilization and redeployment of our scientific resources. Today's big scientific-technical problems cannot be expected to remain forever to challenge the institutions that were mobilized to cope with these problems. The institutions and their sponsoring agencies must inevitably be prepared to move into areas outside their original interests.

The Success of the National Laboratories

Do the large project laboratories merit the immortality as project laboratories they would enjoy if an unending succession of new and important tasks were assigned to them? Or would one do better to shut down project laboratories as they accomplish their tasks and start new ones as new tasks are identified? Both approaches have been tried. The original Metallurgical Laboratory died when the war ended, but it gave birth to two infants that eventually became the Argonne

127

National Laboratory and the Oak Ridge National Laboratory. By contrast, when our country entered the space race in a serious way in 1958, we converted the old NACA laboratories into space laboratories, and in addition built new laboratories.

I admit to prejudice in favor of the large mission-oriented federal laboratories. There are important reasons why these institutions deserve, if not immortality, at least the opportunity to aim for immortality *as project laboratories*. Perhaps the most important justification for this claim to immortality, at least in the institutions I know best, the atomic energy laboratories, is their past success. I shall therefore digress a bit to recount some of the successes of the atomic energy laboratories. I shall omit the many impressive achievements in basic research of these institutions since, for the moment, I am concerned with them as mission-oriented, applied laboratories.

The first major applied success of these laboratories was the establishment of the initial chain reaction at Chicago by Fermi. This was followed by the design and construction of the Hanford plutonium-producing reactors, an enterprise shared by the Du Pont Company and the Metallurgical Laboratory. At the same time, the two major laboratories devoted to separating uranium-isotopes — the SAM laboratories at Columbia in concert with Union Carbide, and the Radiation Laboratory at Berkeley, together with Tennessee Eastman — brought into operation the K-25 Gaseous Diffusion Plant and the Y-12 Electromagnetic Separations Plant at Oak Ridge. Today we look upon these accomplishments with detachment; yet, together with the fashioning of the first nuclear bombs, I think they stand unequaled as examples of what human ingenuity and dedication can accomplish.

From the initial successes in producing Pu^{239} and U^{235} flowed successes in applying nuclear energy to the military arts. First came the Los Alamos atomic bomb, a development that required the perceptive leadership of J. Robert Oppenheimer. In 1953, again at Los Alamos, this time under the general direction of Norris E. Bradbury, the theoretical ideas of Edward Teller and Stanislaw Ulam were translated into a workable hydrogen bomb. At about the same time, the pressurized water power plant for the submarine *Nautilus* was created. It was originally conceived at Oak Ridge under Harold Etherington, and later carried out at Argonne under W. H. Zinn and Etherington with the help of the Westinghouse Corporation, the whole being organized by H. G. Rickover. The years since 1953 have seen extraordinary advances in bomb technology, both at Los Alamos and at Livermore; and in naval propulsion, largely at the "captive" naval reactor laboratories, the Westinghouse Bettis Laboratory and the General Electric Knolls Atomic Power Laboratory.

The laboratories' successes in civilian applications of nuclear energy are no less impressive. Radioisotopes, reduced to large-scale practice at Oak Ridge, are now used throughout science, technology, and medicine. Civilian nuclear power, whose importance I have already stressed, also began in the national laboratories. The earliest concepts of pressurized water reactors were designed during the war at the Metallurgical Laboratory. These ideas were elaborated at Oak Ridge around 1946, and were much influenced by the first high-powered compact water-moderated reactor, the Materials Testing Reactor (MTR). The MTR was carried out as a joint project of Oak Ridge and Argonne. Shippingport, the first commercial American nuclear electric plant, was based on the technology developed for *Nautilus* at Argonne and Westinghouse. Further developments, such as the direct boiling cycle, began at Argonne, and were then elaborated and improved by the private companies, notably General Electric.

Of course there have been failures, such as the aqueous homogeneous reactor which I espoused so strongly, or the nuclear powered airplane. These have been expensive and heartbreaking; but unless we have technical failures as well as technical successes, we are not probing the limits of technology far enough. As long as the entire enterprise is showing healthy progress, we can take some failures in stride. The need for some research environments that encourage technological risk and adventure is one of the justifications for the national laboratories. We need institutions whose survival is not always a pressing issue and which therefore can try the hard as well as the safe projects. An important difference between national laboratories and private laboratories, even those working for the government, is that the national laboratories are buffered from the most acute pressures of survival to a greater degree than are private industrial laboratories. The resulting environment encourages the national laboratories to take greater technological risks than can private, single-purpose laboratories. This environment is an important national asset which I believe must be preserved.

The Trend from Mission to Discipline

The projects and missions that occupy the atomic energy laboratories, though very large and very difficult, are finite. Eventually the projects will be completed, or they will be judged to be so hard that they will be discontinued. In either case the future of the laboratories as project institutions will be in doubt.

An example is civilian atomic power, a field in which our country has made vast strides during the past fifteen years. Economic power based on low-conversion reactors is here. Additional development of converters tends to be incremental, and much of it is carried out privately. As I explained in Part I, the central remaining technical problem is the development of a safe and economical breeder reactor. This has now been recognized as our main order of nuclear business, and the Atomic Energy Commission has formulated a fifteen-year plan aimed at achieving a successful breeder by 1980. Though this is a long time, it is a short time as the history of institutions goes. What then will become of the national laboratories when, say in a decade or so, the breeder reactor has been developed?

One course would be to move away from projects toward components. As long as there is an atomic energy enterprise, and as long as the government retains some responsibility for it, the laboratories will have to investigate a host of smaller matters that are relevant to atomic energy. For example, what limits can be placed on the damage induced in pressure vessel steels by fast neutrons (this becomes increasingly important the longer reactor pressure vessels remain in service); or what can be done to reduce the hazard of a runaway reactor; or how can we shave another 0.1 mill/kwh from the cost of the chemical processing of spent fuel?

Another course would be for the laboratories to stress basic research more heavily. The fraction of a mission-oriented laboratory's effort that should be devoted to basic research is at best a matter of judgment. Inevitably, the bias toward basic research becomes stronger as the preoccupation with projects diminishes. The Chicago Metallurgical Laboratory was focused almost entirely on a single wartime project. Its two daughters, Argonne and Oak Ridge, though both project-oriented, perform relatively much more basic research than did the original Metallurgical Laboratory.[3]

The large laboratories will always be called upon for component development and for basic research. To both of these the laboratories' interdisciplinarity adds unique strength. Moreover, the hierarchical style of organization enables the laboratories to deploy people more aggressively than is possible in, say, a university. As I have already said, at the big national laboratories the whole is more than the sum of its parts: competent scientists in their close interactions mutually

[3]This trend toward basic research is quite noticeable at Harwell, according to its former director, F. A. Vick. See *The Organization of Research Establishments,* Sir John Cockcroft (Ed.), Cambridge University Press, London, England (1965).

131

support each other and produce better results than they could in isolation. The massive attack on the chemistry of the xenon compounds (mostly at Argonne), or the brilliant development of neutron diffraction (at all the AEC National Laboratories), or the concerted development of zonal centrifugation (mostly at Oak Ridge) — these are just a few examples of the great power of the interdisciplinary, hierarchical style of basic research of the national laboratories.

Yet, though this trend from missions (or projects) to components to basic research can be discerned in many federally supported laboratories as their original missions lose their focus, I believe the national interest is not served by allowing the mission-oriented laboratories to lose their mission-orientation as they grow older. This tendency toward loss of mission-orientation is not a remote issue. The mission-oriented federal laboratories tend to be islands in a discipline-oriented academic world. There are always forces brought from the prevailing academic mood that push a laboratory toward disciplinarity. But the applied work of the society must be done, and it is ever so important for these institutions to continue to retain their mission-orientation in spite of the many forces pushing them toward disciplinarity. This means that as missions are accomplished, one must devote deep and serious thought to identification of new missions, perhaps more thought than to any other matter related to management of a large laboratory.

Mobilizing Around New Missions

How do we identify and then mobilize around big, new scientific or technological missions? How does a government agency or a national laboratory decide that now is the time to go, hard, after nuclear energy, or the ICBM, or desalination? A thorough study of this process would be illuminating and well worth the effort put into it. Here I make no pretense to having made such a study. Instead, I shall describe how we mobilized around two problems with which I have been associated — nuclear energy and nuclear desalting — and I shall try to draw some generalizations from these two experiences.

Nuclear energy depended on an accidental finding, the discovery of fission by Hahn and Strassmann in 1938. Immediately upon its discovery, many physicists appreciated its significance: Joliot-Curie in France; Peierls in the United Kingdom; Heisenberg in Germany; Szilard, Fermi, and Wigner in the United States. Of all these distinguished men, I would say that Szilard's role was unique. He had thought about the possibility of nuclear energy as early as 1934; and in 1936 he took out a British Patent (No. 440,023) for a chain reactor based on the $(n,2n)$ reaction in beryllium. The concept of a nuclear chain reaction based on an elementary process in which neutrons are multiplied was therefore already established in his mind. That Szilard was mistaken in his belief that the $Be(n,2n)2He$ reaction was exothermic (it is actually endothermic by 1.67 Mev) should not obscure Szilard's crucial invention: the *idea* of a chain reaction carried out by neutrons. Szilard, of all physicists of the time, was prepared to exploit the discovery of a truly exothermic, neutron-producing reaction — namely, fission. And indeed, Szilard moved with characteristic energy once fission was discovered. It was his idea to persuade Einstein to write his famous letter to President Roosevelt that got the project going in the United States.

My second example is how we have mobilized around nuclear desalting. The obvious importance of a cheap way to desalt the sea had been recognized by Congress when it created in July 1952 the Office of Saline Water in the Department of the Interior. The OSW with admirable dispatch built five demonstration desalting plants, and supported a broad program of basic research relevant to desalting. Yet as late as 1962 no one in authority could see how to reduce the cost of desalted water to the point where desalting could be considered a serious large-scale source of water. It was at this time that R. P. Hammond's views which I described in Part I came to the attention of the President's Science Advisory Committee. Hammond had been asked, as early as 1955, by Senator Clinton P. Anderson of New Mexico, to examine the applicability of nuclear energy to desalting the sea. Hammond soon hit upon his key idea: that bigger meant cheaper, and very big meant very cheap, cheap enough to desalt the sea. Though he recognized this principle quite well by 1956, it took him a full five years of persistent and patient preaching to have the idea accepted by the technical community. But his persistence paid off, and now the world is embarked on a full-scale effort to develop large-scale desalting.

In both of these examples, two essential elements were present. First, there was a single man, or small group of men, who really cared: Szilard in nuclear energy, Hammond in desalting. They focused thinking, not only on a possible solution to the problem but also on the existence, and importance, of the problem to which they had the solution. And second, there was a major technical breakthrough: discovery of nuclear fission in the one case, and recognition of the advantage of size in the other.

Both elements seem usually to be needed for the successful mobilization of large, ponderous scientific organizations around the big new problems. Without a clear point of departure, without some kind of breakthrough, the most one can do is mobilize one's resources to try to find a way of achieving a desirable aim. Before Hammond's idea of size emerged, there was much good work done on desalting, but it was diffuse in its aims. Once the initial technical breakthrough has been achieved, exploiting the breakthrough efficiently becomes a tactical matter.

The great man — the single individual who sees both what can be done and what must be done — is almost as important as is the great idea. For the hand of any scientific bureaucracy is always heavy. To force redeployment of a large array of scientific resources around a new mission requires energy, skill, and dedication of an unusual order. A dedicated and single-minded man — a Rickover or a Szilard — often makes this redeployment possible when in his absence it is all but impossible.

Within any large laboratory there are internal pressures that resist taking on big new problems that are very different from those the laboratory has been doing. Most scientists like to do what they have already done. Relatively few have the intellectual strength and confidence to do, of their own accord, what Fermi once told me he did: change fields every five years. Yet in many cases the actual change in what the individual does when the institution's objectives change may not be so great after all. To a chemist working on the corrosion of aluminum caused by hot water, it makes little difference whether the water comes from a reactor or a sea-water still. The specialties needed to attack desalting are about the same as those needed to attack reactor development: chemistry, chemical engineering, physics, metallurgy, mechanical engineering. Since the AEC's National Laboratories are so well equipped with experts in many disciplines, they would be particularly able to redeploy around new problems of this sort.

A more serious obstacle to redeployment is the parceling of our government's work among the various government agencies. Each government agency is responsible for achieving certain ends, usually ends that are suggested by the names of the agencies: the Department of Defense, the Department of Health, Education, and Welfare, the Space Agency, the Atomic Energy Commission, and so on. The laboratories that belong to each of the agencies naturally work on missions that fall within the responsibility of the sponsoring agency. This imposes a rigidity on the system. When a government laboratory finishes a project, it cannot ask, What is the most important national problem, as seen from the widest possible viewpoint, to which our talents can be put? Rather, the lab-

135

oratory must ask, What is the most important problem, *coming within the purview of our sponsoring agency,* to which we should next turn? This narrower set of problems may not be as important to the nation as are other problems which are the responsibility of a different agency, but which the laboratory may be equipped to handle. Such rigidity reduces the efficiency with which we deploy our federal scientific apparatus. What we need are government, as well as government agency, laboratories — true national laboratories, as well as national atomic energy laboratories, national space laboratories, national defense laboratories.

This possibility was recognized by the Atomic Energy Commission in its February 1960 Report to the Joint Committee on Atomic Energy on "The Future Role of the Atomic Energy Commission Laboratories." We find on page 7, Volume 1, of this report the statement, "the Commission will utilize these laboratories and their staffs for other urgent tasks or projects of importance to the nation." This trend toward wider deployment is discernible both in the United States and abroad. Oak Ridge now works on desalting for the Office of Saline Water; on chemical carcinogenesis for the National Institutes of Health; on civil defense for the Office of Civil Defense. Harwell in England and Saclay in France have both begun work on desalting. I believe this is proper and that as new important missions are identified, we should cross bureaucratic lines, both internal and external, and deploy our resources where they can best be used.

Obsolescence of mission is likely to pose a more acute problem if the agency is organized around a technology rather than around a public purpose. This is perhaps best illustrated by the Atomic Energy Commission, which was set up to exploit a scientific discovery: nuclear fission. It seeks to provide us with energy, but only through the technology for which it is, by law, responsible — that is, through nuclear processes. At the time the Commission was established, it seemed to be most expedient to get on with the job of nuclear energy by establishing an agency specifically for this purpose. But now the AEC's one-sidedness leads sometimes to unseemly rivalry between fossil fuel, represented by the Department of the Interior, and atomic fuel, represented by the Commission. A more rational approach would be to establish a government agency with responsibility for energy — atomic, fossil, solar. Of course, if my predictions about nuclear energy displacing all other forms of energy are realized, then the Atomic Energy Commission will become a de facto Energy Commission.

A crisis of mission will face our space program during the middle seventies. The present scale and pace of our space enterprise is determined by our commitment to send a man to the moon early in 1970. Now that this date is approaching, grave questions arise as to the next step. Should we go on to Mars; or should the space program revert to the scale it enjoyed before we decided to send a manned expedition to the moon? The existence of a massive space establishment, including many fine laboratories, surely will tend to weigh the balance strongly toward going to Mars after we have reached the moon. Yet I would hope that when the time for this far-reaching decision comes, we shall have ready for this great scientific-technical resource many other technical challenges that bear more directly on our human well-being than does a manned trip to Mars.

The Problem of Big Problems

Are there new broad areas where the next generation of Big Problems are likely to lie? Certainly there are; I shall enumerate only some obvious ones, most of which arise from attempts to resolve the Malthusian dilemmas I mentioned in Part I.

I have spoken many times of the central importance that I attach to the development of the economical breeder. No new technical breakthrough is needed here; the successful

137

breeder ought to come from relatively straightforward application of known principles. I believe therefore that this task should be completed comfortably by 1980, the date set by the Atomic Energy Commission. I view the matter as so important that I would like to see much more of our atomic energy enterprise devoted to the breeder than now goes into it. And, until the breeder has been achieved, the national atomic energy laboratories have not really accomplished their primary mission, nor can one properly speak of their having "worked themselves out of a job."

Beyond the breeder, I see many possibilities. There is thermonuclear energy. Though the outlook for success fluctuates, I see nothing wrong with pushing controlled thermonuclear fusion at a reasonable rate, just as long as the scientists continue to find points of departure. When and if a breakthrough is found, the effort would be very sharply expanded.

I believe the nuclear energy laboratories ought gradually to investigate conversion of energy to the other necessities of life, a possibility I mentioned in Part I. Desalting is an example of such a use of energy. But the other necessities — fertilizer; liquid fuel from coal or, more generally, mobile, nonpolluting secondary energy sources, particularly for automobiles; and metals from low-grade ores — are matters that are almost ripe for large-scale investigation by the big laboratories even without any great new inventions. Certainly the laboratories themselves are well equipped to attack any of these problems.

Beyond these fairly obvious technological matters, I am much impressed with the promise of very large-scale, mission-oriented work in the biomedical sciences. As I have stated in Part III, I believe biomedical research in the style of Big Science is coming, and that it will require multidisciplinary institutes. I should think that within a decade or so several of the large government laboratories that are now predominantly concerned with the physical sciences and engineering will be called upon to help in the massive attack on human disease, especially the ailments traceable to subtle alterations of our environment.

Most of the foregoing Big Problems — nuclear breeding, thermonuclear energy, desalting, conversion of energy into fertilizer and into other means of subsistence — are primarily technological problems. To be sure, each has economic components; a breeder must be economical if it is to be successful. Yet the problems can be stated rather precisely, and their technical solutions ought to be easily recognized. For this reason these problems can be, and some are being, handled within the big technologically oriented national laboratories.

In the last half dozen years, we have begun to recognize entire new classes of problems that have a much stronger social component. One of the first to come to some focus was the problem of pollution; this was dramatized in 1962 by Rachel Carson,[4] and it has since been the subject of a major study by the President's Science Advisory Committee.[5] Other problems of this sort that, like pollution, arise from the expansion of our industrialized population are urban renewal and mass transportation, not to speak of control of population.

Another great problem with deeply ramified social, political, and technical implications is civil defense. Since President Kennedy set up the Office of Civil Defense, research on civil defense has been carried out at a reasonable rate. But the whole matter has been brought much more sharply to a head by Eugene P. Wigner, who argues that the present disparity between offense and defense presents an intolerable temptation to an aggressor. Civil defense, as a means of redressing this imbalance, is therefore viewed by Wigner as a crucial element in staving off thermonuclear war, and therefore a

[4]Rachel Carson, *Silent Spring,* Houghton Mifflin Company, Boston, Mass. (1962). I touched upon this matter in Part I, page 30.

[5]"Restoring the Quality of Our Environment," Report of the Environmental Pollution Panel, President's Science Advisory Committee, U.S. Government Printing Office, Washington, D.C. (November 1965).

proper technical question around which to deploy on a large scale.

These newly emerging Big Problems have several common features. Unlike nuclear energy, where a major discovery focused our attention upon the problem of energy, the order here is reversed. Urgent social problems are first recognized as problems, and we then mobilize around them even though we don't have any very clear idea as to how to achieve their solutions. Being social problems, they are hard to delineate precisely. When did we realize our air was polluted or our cities so decayed as to require major effort aimed at cleaning the air or renewing the cities? The solutions, when they come, will be equally hard to identify. When is our air "clean," or when are our cities "renewed," or when is our civil defense "adequate"? These problems involve countless decisions by many individuals. The possibility of all-pervading, technically based solutions is therefore usually discounted.

It is for this reason, I believe, that these new social problems do not naturally gravitate to the existing federal laboratories. The latter are technologically oriented, and have little tradition in the social sciences. Yet attempts to redeploy some of these laboratories around problems with strong social components are under way. For example, I have mentioned that the Oak Ridge National Laboratory has undertaken a study of civil defense. Some of the questions are clearly social: Why does civil defense evoke such hostile response in many quarters, whereas increased commitment to retaliatory weapons draws little attention? Or, how will populations behave in shelters? But many of the questions are technical: the structural design of shelters, resistance to fire storm, assessment of the ecological balance after a thermonuclear war. Since both the natural and behavioral sciences are involved, it seemed just as logical to add social scientists to the existing core of natural scientists at Oak Ridge as the other way around. The resulting mixture of behavioral and natural scientists has, I believe, been working effectively on civil defense even though the environment in which they work, a large technically oriented laboratory, may at first seem to be an unlikely one for such an enterprise.

There is a possibility that the technologically oriented research institutions may contribute to an unexpected degree to the resolution of problems that now seem to be primarily social. I refer to the possibility of devising "cheap technological fixes" that afford shortcuts to resolution of social problems. One of the simplest, and possibly most important, such technological fixes, is the Gräffenberg ring. Before the invention of the ring, birth control, because it was awkward, required strong individual motivation. The ring, being a simple, one-shot method, requires much weaker individual motivation. It provides a promising technological path to the achievement of birth control without having first to solve the infinitely more difficult problem of strongly motivating people to have fewer children.

Another quite unexpected possibility, which I have alluded to in Part I, is the use of nuclear energy as a means of reducing air pollution. Our atmosphere is being polluted by the effluents from countless individual fossil-fueled energy sources — power plants, automobiles, space heaters. We can clean our air if we pass laws that require people to add afterburners to their cars or better precipitators to their power plants. Since all such measures are costly, there is strong resistance to taking the many separate steps needed to clean up the environment. But if our energy system became all nuclear, and particularly if we could invent a lightweight, powerful storage battery or fuel cell to power automobiles, some of our worst air pollutants would disappear. Moreover, if the economic projections of nuclear energy I made in Part I come to pass, there will be economic incentive to use this cleaner energy source. Energy would not only be cheaper; it would be cleaner. And since it is cheaper, people will use it, not because a law requires them to, but because it is to their individual economic advantage to do so. In this sense, a technological advance — cheap and clean nuclear energy — would bypass the necessity for deliberately inducing social change by legislation.

The foregoing are a few of the Big Problems set, basically, by the growing pressure of an expanding population. It is heartening to see our ponderous scientific bureaucracy gradually address itself to these issues. The federal laboratories, insofar as they can become involved in the attempts of society to deal with these Big Problems, will to this extent remain viable and relevant to the society that supports them.

Institutions and Society

My concern about the long-term relevance of the national laboratories to the society that supports them differs little from similar concern, voiced, for example, so delightfully by C. Northcote Parkinson,[6] for the ultimate relevance of every institution of which our society is composed. Every institution within a society originally serves a purpose recognized and defined by the society. When the institution is young, it is usually responsive to those purposes. Its health and prosperity are co-extensive with the health and prosperity of the society in which it is embedded. But gradually the institution acquires its own, internally generated imperatives: aspirations for greatness, for power, for immortality. It acquires loyalties to its own viewpoints and prejudices. The naval officer believes national defense is best maintained through command of the sea, the air force officer, through command of the air, the army officer, through command of the land. How do societies reconcile the inherent aims and aspirations of the institution with the aims and aspirations of the society; how do they make the institution serve the society rather than the other way around?

In coping with this intrinsic conflict between the part and

[6]C. Northcote Parkinson, *Parkinson's Law,* Houghton Mifflin Company, Boston, Mass. (1957).

142

the whole, between the institution and the society, every society uses the cabinet system. The President brings into the highest councils of government the heads of the agencies; the agency head brings into his council, department heads; the department head, section chiefs; and so on. The purpose of a cabinet is twofold: first, the cabinet gives the leader advice, and second, it gives the department heads a sense of identification with larger purposes that transcend individual departments.

These two functions of the advisory council are often in conflict. Department heads are apt to temper their advice with too strong a concern for the health of their own department. The leader therefore naturally tends to stack his cabinet with staff whose loyalties to him and to the larger unit are undiluted by loyalties to the smaller unit. If the advisers happen to be brighter than the department heads, the leader thereby gets sounder advice and formulates sounder policies, but in so doing he weakens his ability to implement his policies. The inertia of a balky department is monumental. To force it to do what it should is difficult at best; it is doubly so when the department has had nothing to say about formulation of the policy it is asked to implement.

In applying these observations to the problem of laboratory mission, I would urge that laboratory managements be encouraged and enabled to view their purpose from vantage points larger than that of their laboratory. I, for example, have always valued my term on the President's Science Advisory Committee (PSAC), not so much for what I may have contributed to the committee's deliberations, but for the insights it gave me as to the proper role of the large federal laboratory I direct. I believe Oak Ridge better serves the nation because my viewpoint was broadened by my term on PSAC. I would therefore urge that our scientific government give to those who help operate federal research establishments more responsibility for formulating policies which ultimately involve their institutions and to which their institutions can contribute. I would urge, for example, that the advisory apparatus of the scientific government, the President's Science Advisory Committee, the General Advisory Committee of the Atomic Energy Commission, the advisory councils of the National Institutes of Health, be more liberally peppered with laboratory directors. The director may be, superficially, in conflict of interest, but I think our national scientific resources would be readier to work on what is most in the national interest if the institutions and their heads were more strongly identified with the larger purposes they are supposed to serve.

For the purposes of an institution, especially a publicly supported institution, can hardly be determined from within the institution. Just as I argued in Part III that the value, as contrasted to the methodology, of a scientific discipline cannot be established entirely from within the discipline, so the purpose, as opposed to the expertise, of a national laboratory cannot be established from within the framework of the laboratory itself. The management of the institution must therefore retain a sensitivity to the needs of the agency, the department, or the society which it serves. To the extent that it succeeds in retaining this sensitivity — that is, to the extent that the national laboratory remains deployed against the Big Problems, and only to that extent, will these institutions of Big Science and Technology continue to merit the support society now so generously gives them.

2. Universities and Disciplines[7]

Our society is mission-oriented. Its mission is resolution of problems arising from social, technical, and psychological conflicts and pressures. Since these problems are not generated within any single intellectual discipline, their resolution is not to be found within a single discipline. Society's standards of achievement are set pragmatically: what works is excellent, whether or not it falls into a neatly classified discipline. In society the nonspecialist and synthesizer is king.

The university by contrast is discipline-oriented. Its viewpoint is the sum of the viewpoints of the separate, traditional disciplines that constitute it. The problems it deals with are, by and large, problems generated and solved within the disciplines themselves. The university's standards of excellence are set by and within the disciplines. What deepens our understanding of a discipline is excellent. In the university the specialist and analyst is king.

The structure of the discipline-oriented university and the structure of the mission-oriented society tend to be incongruent. Moreover, as the disciplines making up the university become more complex and elaborate in response to their own internal logic, the discrepancy between the university and society grows. The university becomes more remote; its con-

[7]Based on an article "But Is the Teacher Also a Citizen?", *Science 149*, 601–606 (August 6, 1965).

nection with society weakens; ultimately it could become irrelevant. The growth of this discrepancy appears to me to be a central problem in the relation between the university and society. It poses major difficulties for the university professor, especially in the natural sciences, who views broadly his responsibility as a citizen.

Harvey Brooks, Dean of Engineering and Applied Physics at Harvard University, put the matter with his usual incisiveness:[8]

"The . . . issue is the relationship between science and technology in education. The original concept of an engineering school, as of a medical school, was an association of practitioners who used the benefit of their varied experience to teach young people. This tradition is somewhat maintained to this day in the field of architecture, but in both medicine and engineering the importance of the underlying sciences has become so great that medical and engineering faculties are increasingly populated with basic scientists who do research or teaching in sciences which are relevant to but by no means identical with the practice of medicine or engineering. The old form of teaching primarily by practicing physicians or engineers was found wanting because practical knowledge was too rapidly being made obsolete by new scientific developments which could not be fully absorbed or appreciated by the mature practitioner. Yet in the process something of the spirit and attitude of the skilled practitioner was lost, particularly *his willingness to deal with problems whole rather than in terms of the individual contributing disciplines*. . . . In medicine this problem has been partially met by the teaching hospital, but in engineering the analog of the teaching hospital is the big engineering development laboratory in industry. How, then, is the spirit of applied science and engineering to be retained in engineering educa-

8"The Dean's Report, 1963–1964," *Harvard Engineers and Scientists Bulletin 45*, p. 6 (December 1964).

tion? The intellectual foundations of engineering lie increasingly in the basic sciences; inevitably engineering faculties will contain large numbers of people whose way of thinking is more akin to that of the scientist than the engineer. It is these people who will develop many of the techniques which will be used by the engineer of the future. And it is their knowledge, not that of the current engineer, which the student will be using ten years from now. The reconciliation of these two necessary attitudes of mind in the process of engineering education is the central dilemma of the field today."

The Trend Toward Purity

Though Brooks's critique is directed mainly at the engineering school, what he says has wider relevance. The university's disciplinary viewpoint and even organization create many points of tension between the university and the society in which it is embedded.

One is the tendency toward increasing purity, especially in the sciences and most notably in mathematics. I would measure "purity" of a branch of science by the degree to which the phenomena studied are of intrinsic interest to that science or are of extrinsic interest. In the first instance, the science is purer; in the second, where the motivation is to understand phenomena which lie outside the branch, the science is less pure. Thus I would divide science into pure or intrinsically motivated, and applied or — more broadly — extrinsically motivated. For example, applied science (in the usual sense of the term) seeks to clarify some aspect of, say, engineering or medicine. We study the chemistry of molten fluorides at Oak Ridge because we wish to build a reactor that uses molten fluorides; or we study certain viruses because these viruses are implicated in certain kinds of leukemia.

Extrinsically motivated science also includes those sciences that are pursued to deepen our understanding of some other

147

branch of pure science. For example, those parts of nuclear physics that are studied primarily to elucidate the origin of the elements rather than the structure of the nucleus would, in my usage, be termed extrinsically motivated. On the other hand, the study of elementary particles, originally motivated by our desire to understand the nuclear force, now develops with an internal force dictated by the intrinsic interest and beauty of the phenomena occurring at very high energy. I would therefore call elementary particle physics pure. Of course it is in the nature of pure science that the light it eventually will shed on other branches of science or technology is to some degree unpredictable; yet at any given time I believe one can often make a judgment of relevance on the basis of the motivation of those practicing the science. Thus, as I pointed out in Part III, many nuclear physicists who measure capture cross sections make no bones about their primary motivation; it is to help the astrophysicist better understand stellar nucleosynthesis, rather than to help themselves better understand the nucleus.

At its inception nearly every science is extrinsically motivated — that is, it seeks to explain questions that were originally part of some other branch of human interest, usually, though by no means always, some practical matter. Mathematics originated because men had to measure, weigh, and count to maintain an organized economic system. The study of thermodynamics started from Carnot's interest in steam engines. Pasteur's science of bacteriology began when he tried to prevent French beer and wine manufacturers' products from turning sour. Group theory was invented as a means of studying the properties of algebraic equations. So to speak, nearly every pure science starts as an applied, or at least as an extrinsically motivated, science.

And, indeed, in previous generations the distinction between pure and applied science, and between pure and applied scientists, was less pronounced than it is today. The three greatest pure mathematicians — Archimedes, Newton, and Gauss — were also great applied mathematicians; to these one can add the three greatest pure mathematicians of the twentieth century — Poincaré, Hilbert, and von Neumann — each of whom was also a great applied mathematician. Pasteur, the founder of bacteriology, was an applied scientist. Lord Kelvin was equally at home in applied and basic physics. Similarly, the distinction between theoretical and experimental science was much less sharp two generations ago than it now is. Maxwell did experiments and also constructed theories.

But daughter sciences, once they bud off the stalk of the parent science, acquire a separate existence, grow, and luxuriate. In the process these offspring generally become purer and narrower. The parent stalk had closer roots in the original questions posed by some urgent need: in chemistry, the need to extract metals, or to find the elixir of life; in mechanics, to build more accurate missiles; in astronomy, to predict the seasons. But today many pressures compel the daughter science to become purer, especially when the science is pursued within the university.

To understand how this comes about, I point out that every scientist or, for that matter, any intellectual creator, in plying his trade, tries to choose for himself problems that are both soluble and important. The importance of a problem is judged by the scientist by the breadth of added understanding its solution affords. The discovery of the second law of thermodynamics was important because it organized so many otherwise disjointed elements of physics and chemistry. Its discovery was much more important than, say, the discovery that light reflected at the Brewster angle is completely polarized, since the latter discovery affects a much narrower segment of related science or technology. The important questions often tend to be posed as much from without as from within a given narrow field of inquiry: in the case of the second law, Carnot was seeking to clarify fundamental questions in heat engineering. The solution of an important problem tends to reinforce the relation between a scientific discipline and the disciplines to which it is related. In this sense the important questions are broad; they tend to be extrinsically motivated.

Unfortunately, the important questions are often the most intractable ones, and therefore most of science is concerned with soluble problems, not important problems. We do not know how to create a controlled thermonuclear plasma; we therefore study aspects of plasmas that are tractable rather than necessarily relevant in the hope that our added general knowledge will eventually help us make progress toward the goal of controlled fusion. But in the process the science of plasma physics becomes purer. So, in general, the strategy of pure science is always to deal with soluble problems which, by their nature, tend to be narrow in impact. The important problems are skirted until enough soluble problems have been solved to permit a successful attack on the important problems.

The social structure and the purpose of the university accentuate the pressure toward purity. For the university's purpose is not to solve problems that are set from outside a discipline. Its purpose is to create and encourage the intellectual life per se. If a scientific discipline sets off on an independent course, separate from its original applied parent, it tends in the university toward greater purity and remoteness simply because there are few countervailing pressures there. In the university it is improper to ask of the scientist, What is the relevance of what you are doing to the rest of the world or even to the rest of science? The acceptable question is, What do your scientific peers, who view your work with the same intellectual prejudices as you, think of your work?

The process leading toward greater purity and remoteness was described with exquisite perception by John von Neumann, though it had been discussed previously by David Hilbert.[9] I have already quoted von Neumann's views in

[9]Cf. Lecture before the Mathematical Congress, Paris, 1900: "In

Part III, and I shall therefore repeat only a few sentences for emphasis: "As a mathematical discipline travels far from its empirical source . . . there is a grave danger . . . that the stream so far from its source will separate into a multitude of insignificant branches, and that the discipline will become a disorganized mass of details and complexities . . . whenever this stage is reached, the only remedy seems to me to be the rejuvenating return to the source: the re-injection of more or less directly empirical ideas. I am convinced that this was a necessary condition to conserve the freshness and vitality of the subject and that this will remain equally true in the future."

Von Neumann's plea for greater unity in the mathematical sciences has been taken up by others, notably Mark Kac and Richard Courant, who see danger in the trend toward super-purity, abstractness, and remoteness. Kac[10] speaks of the professional purist in mathematics; Courant[11] speaks of the "isolation that threatens every pursuit of science — certainly very much the pursuit of mathematics — this isolation can be very stifling." The trend toward isolation that has marked modern mathematics seems to me to have invaded the empirical sciences, and possibly even the social sciences, and for the same reasons. For example, the nuclear structure physicist today concerns himself with subtler, more delicate

---

the meantime, while the generating force of the pure thought acts, the external world enters again and again; it forces upon us new questions through the actual phenomena; it opens new topics of mathematics, and while we try to incorporate these new topics into the realm of pure thought, we find often the answers to old unsolved problems, and advance this way best the old theories. Upon this repeating and changing interplay between thought and experience are based, as it seems to me, the many and surprising analogies and that apparently 'prestabilized harmony' which the mathematicians so often notice in the questions, methods, and concepts of different fields of science."

[10]"Mathematics: Its Trends and Its Tensions," Occasional Paper No. 10, p. 8, The Rockefeller Institute Press, New York (1961).

[11]"The Working Mathematician," *International Science and Technology*, 83 (March 1965).

questions about nuclear structure than he did twenty years ago. And just because the questions are subtler, and more detailed, they tend to have less relevance to the fields of science and technology that surround nuclear structure physics. The language of the nuclear structure physicist becomes more sophisticated, his techniques more specialized. His ability to communicate with his colleagues in surrounding fields becomes impaired; and insofar as what he studies becomes of less relevance to the fields in which his own field is embedded, his own field becomes purer.

The Denial of Science as Codifier

Science traditionally has two aspects: it is on the one hand a technique for acquiring new knowledge; it is on the other hand a means for organizing and codifying existing knowledge, and therefore a tool for application. Both aspects of science are valid. The discovery of SU_n symmetry does not in the slightest detract from the importance of the second law of thermodynamics. This law, with its enormous power as an organizing principle for much of existing chemistry, though discovered more than a century ago, is as much science as is the *search* for new unitary symmetries.

The modern university tends to emphasize science as search at the expense of science as codification, and for many of the same reasons it drives science toward fragmentation and purity. As I mentioned in Part II, the codified parts of science are often most useful in the neighboring sciences, or in engineering applications, not in the science in which the codification originally took place. X-ray crystal analysis sprang up in physics; most x-ray crystallographers now work as chemists, metallurgists, or even biologists. The university's disciplinarity, its tendency to deal with pure problems that are intrinsically motivated, reduces its concern for science as codification; such science has already been bypassed by the researcher in the field.

152

The pressure to do research rather than to teach accentuates the denial of science as codification. Much has been said about the conflict between research and teaching in the large university. As I see it, at least part of the conflict amounts to a philosophic judgment whether science is the search for new knowledge or the organizer of existing knowledge. In emphasizing research at the expense of teaching one is implicitly valuing the one above the other.

One by-product of this trend is the waning of the tradition of scientific scholarship. As our sciences become more and more fragmented and narrowly specialized, and as their connection with earlier, more general phases of science weakens, the relevance of what came before to the pursuit of current research decreases. For example, it is possible to carry out research on elementary particles without knowing much about nuclear structure. The taste for knowing the historical origin and development of a science wanes, partly because such knowledge is unnecessary for prosecuting current research and partly because one has too little time and energy left over after learning what is needed to do the research at hand.

Implications: For Education

These two tendencies — toward purity and fragmentation as opposed to application and interdisciplinarity, and toward research and away from scholarship — seem to me to portend trouble in the relation between the university and society. First, I speak about the great curriculum reforms, especially in the sciences. These reforms started in the high schools but have now been extended, particularly in mathematics, downward to the grade schools, and in many instances upward to the colleges. They are relevant to my discussion, because the reforms have been instigated by the university, and they certainly reflect the intellectual spirit of the university. With certain of the aims of the curriculum reforms, one can have no quarrel. The new curricula try hard to be interesting, and in this I think they succeed. Some, like the new curriculum in biology, seem to me to inject a welcome note of unity into the subject. The curricula generally demand more effort and present more challenge than the old. But, insofar as many of the new curricula have been captured by university scientists and mathematicians of narrowly puristic outlook, insofar as some of the curricula reflect deplorable fragmentation and abstraction, especially of mathematics, insofar as most of the curricula deny science as codification in favor of science as search, I consider them to be dangerous.

The danger I worry about was brought home to me by a distinguished physics professor. According to him, the mathematics department at his university no longer teaches the kind of calculus course which develops power and skill in handling simple integration. Such skills are apparently too lowbrow, and in any event are no longer needed by one who wishes to pursue a career as a research mathematician. As a result, many physics students are unable to do the mathematics that still is important to physics, even if not to mathematics. This physics professor has therefore written a book on calculus which presents the traditional parts of the subject that have been bypassed by the professionals. I think this anecdote illustrates both what is wrong with, and what might be done to remedy, the situation. The professional purists, representing the spirit of the fragmented, research-oriented university, took over the curriculum reforms, and by their diligence and aggressiveness, created puristic monsters.[12] But education at the elementary level of a field is too important to be left entirely to the professionals in that field, especially

[12]To those who are skeptical of this criticism, I recommend the report "Goals for School Mathematics," Houghton Mifflin Company, Boston (1963), that purports to describe a possible mathematics curriculum for the 1990's. One proposal (fortunately not accepted by all of the members of the study group) suggested that complex numbers be introduced into the eighth grade as "residue classes on the field of polynomials modulo $(x^2 + 1)$"! This report received the sanction of the then U.S. Commissioner of Education, F. Keppel!

if the professionals are themselves too narrowly specialized in outlook. Instead, curriculum reforms should be strongly influenced by disciplines bordering the discipline being reformed. The mathematics curriculum should receive strong cues from the empirical sciences and from engineering; the physics curriculum, from engineering as well as from the neighboring sciences, and so on. There is nothing wrong with physics professors writing calculus books, or engineering professors writing physics texts, as long as the physics professor knows calculus or the engineering professor knows physics. And, indeed, seeds of the counterrevolution in curriculum reform seem to be sprouting. In mathematics a counterrevolution seems to be taking place; for example, a group of seventy-five leading American mathematicians stated: "to offer such subjects to all students as could interest only the small minority of prospective mathematicians is wasteful and amounts to ignoring the needs of the scientific community and of society as a whole."[13] And the American Council for Curricular Evaluation has been organized to maintain "the intellectual integrity of our schools" — that is, to scrutinize some of our newer curriculum reforms.

Related to the trend toward purity in curriculum reform is the relatively poor place of applied science in the universities. This matter has been emphasized by Edward Teller,[14] and I have already mentioned it in Part III. Teller points out that most of the money our government spends for research and development goes for applied research; yet most of the prestige and emphasis in the university goes to basic science. Hans Bethe, in speaking of the social responsibility of the scientist, has also noted this denigration of the applied sciences in the university. He exhorts the university scientist to

[13]L. V. Ahlfors, et al., "Mathematics Curriculum of the High School," *Am. Math. Monthly 69,* 189–193 (1962).
[14]Cf. Teller's paper in "Basic Research and National Goals," 257–266, A Report to the House Committee on Science and Astronautics, submitted by the National Academy of Sciences, Washington, D.C. (March 1965).

155

overcome his prejudice against application and especially urges him, as part of his social responsibility, to reaffirm the dignity of applied science.[15]

Implications: For Government

What are the implications of these trends for government and society? Our society increasingly is a product of the university. As the university degree becomes more and more common — it may be nearly as common, eventually, as a high school diploma now is — the outlook and point of view of our society and of our government become the outlook and point of view of the university.

I want to make perfectly clear that on balance I believe this to be enormously good. The university is rational, and its outlook is basically tolerant and knowledgeable. For example, I believe our whole enlightenment in race relations would be unthinkable if anthropological and psychological doctrines, developed largely in the university, had not penetrated society as a whole. One must never forget that the Supreme Court, in justifying its 1954 decision on school desegregation, invoked a psychological doctrine (psychic damage to the segregated child) that catches the spirit of and was certainly nurtured by the university.

But my purpose is to point out the dangers to government, to society, and to the university that lie in the latter's narrow disciplinarity. Thus the university's picture of science as research and denial of science as codification or as a tool for application deadens its taste for action. Let me illustrate with the views expressed in the report "Growth of World Population," released by the National Academy of Sciences in 1963, to which I subscribed at that time.[16] The report con-

[15]"The Social Responsibilities of Scientists and Engineers," a lecture at Cornell University, November 6, 1963, and published subsequently in *The Cornell Engineer* and excerpted in *SSRS Newsletter*, p. 1 (February 1964).
[16]"Growth of World Population," Committee on Science and

cluded that the over-all task was to achieve "universal acceptance of the desirability of planning and controlling family size." The report then made four major recommendations, which I paraphrase:

1. Support graduate and postdoctoral training in demography.
2. Expand research laboratories for scientific investigation of human reproduction.
3. Cooperate in international studies of voluntary fertility regulation.
4. Train more administrators of family planning.

No one concerned with the population problem can take issue with any of these recommendations. Of course we need more research and more studies, as well as more administrators. But such recommendations are, it now seems to me, tangential to the main issue. They substitute *research about* the problem of family planning for *action on* the problem. Complicated social problems such as control of family planning must be attacked with the information at hand even as we learn more about these problems. And, indeed, the distinguished biochemist William D. McElroy, who chaired the panel that issued the report on population, said recently, "Although I am still in full agreement with these recommendations, I think the time has come when we must move ahead even without the additional biological knowledge."[17]

Nor is this instance an isolated one. Panels that advise government, especially on matters having scientific implications (and what affair of government these days does not?), are usually dominated by university people, especially those active in research. What is more natural than to recommend

Public Policy, National Academy of Sciences, U.S. Publication 1091, Washington, D.C. (April 1963).

[17]"The Need for Action," *Population,* Panel discussion held at the Pan American Union, p. 43, Planned Parenthood/World Population, New York (1964).

157

more research as a kind of magical talisman that will solve profound and complex social problems?[18] I am therefore much impressed with the contrast between the recent study on heart, cancer, and stroke,[19] which proposed specific concrete action on the basis of the knowledge at hand, and the many other studies, such as the NAS study on population, which display an inclination to study rather than to do.

Even the choice of what things our government decides to spend its research funds on is now deeply influenced by the puristic university. In earlier, and simpler, times the government's attitude toward science was unsophisticated and inexpensive. First, the nonscientific goals of the society were ascertained by the political process; these goals by and large transcended the goals of the university. Thus, we had long since decided that national defense was a necessary goal; or good public health; or better navigation; or adequate physical and chemical standards. We then decided to support the science that scientists believed would help achieve these goals. How much we spent on the relevant science was determined by how important we regarded the goals themselves to be, and this was a political decision. It is true that in recent years we have become very relaxed over how relevant a science need be to warrant support; nevertheless, the mission-oriented agencies support basic science per se largely as a justified overhead expense charged against achievement of the over-all mission.

The current active debate on scientific priorities bespeaks a change in our viewpoint. Whereas in previous times government support of science was justified by its contribution to the achievement of some nonscientific end, we seem now to have accepted the view that science deserves large support

[18]I have already touched upon this point in Part III. Again I repeat that the "cheap technological fix" may often be very useful in dealing with social issues.

[19]"A National Program to Conquer Heart Disease, Cancer and Stroke," President's Commission on Heart, Cancer and Stroke, U.S. Government Printing Office, Washington, D.C. (1965).

solely for its own sake; with this development no scientist can quarrel. However, to my mind, the same professionally puristic viewpoint that has captured the elementary mathematics curriculum seems to prevail in the present debate on scientific priority. The debate at the moment centers on the support given high-energy physics relative to that given other fields of science. Now high-energy physics is at once the most elegant and, in a sense, the most fascinating branch of physics. The new unitary symmetries are beautiful to behold and astonishingly unexpected. The high-energy physicists themselves are brilliant and dedicated. Because the field is rich and exciting it certainly deserves support. I cannot, however, understand the argument that high-energy physics commands an *urgency* of support simply because, as Robert Oppenheimer puts it, it is "the conviction of those who are in it that, without further penetration into the realm of the very small, the agony may this time not end in a triumph of human reason."[20] The agony Oppenheimer refers to is surely not shared by all of society, nor even by all scientists. The question is why the intellectual agony of this generation of physicists needs to be relieved as quickly as possible rather than being resolved, at a slower pace, by succeeding generations.

The emergence of high-energy physics among our country's highest-priority basic scientific enterprises is a manifestation both of the university's deification of purity in science and of its influence on what our society does. High-energy physics is the purest branch of physics. In the university community it towers above most sciences in prestige and in the caliber of the students it attracts. That it should be placed so high on our society's list of things to be done attests at once to the pervasiveness of the university's influence on the society and to the way in which fragmentation and concern for disciplinary purity of the university, when imposed on the mis-

[20]*Nature of Matter,* Luke C. L. Yuan (Ed.), p. 5, Brookhaven National Laboratory, Upton, New York (January 1965).

sion-oriented society, divert the society from its real goals. Our society is not a university; the goals of our society are not the same as the goals of the fragmented and discipline-oriented university. It is hardly acceptable for the university to persuade the society that at this stage in history the university's own intellectual goals and aspirations — remote, pure, and fragmented — deserve the highest place among the goals of the society.

Recapitulation: The Embeddedness of Values

My views have been a fugue on a single theme, a theme that recurs throughout these essays. I began by pointing out that the university and society are incongruent in that the university is discipline-oriented and fragmented, the society, mission-oriented and whole. I tried to show how the ecology of the discipline-oriented university encourages the rise of purism and specialization and the denial of scholarship and application in science. I then argued that these trends in the universities are affecting our elementary curriculum; are giving us poorer people to get on with the applied work of the day; are substituting research for action; and are tending to impose upon society the scientific values of the fragmented university.

In every one of these trends I discern the same underlying issue: a failure to realize that no judgment of the relative value of a universe can be made from the narrow base of that universe. Values are established from without a universe of discourse; means are established from within. Thus, our science tends to become more fragmented and more narrowly puristic because its practitioners, harried as they are by the social pressures of the university community, have little time or inclination to view what they do from a universe other than their own. They impose upon the elementary curriculum their narrowly disciplinary points of view, which place greater value on the frontiers of a field than on its tradition, and they try to put across what seems important to them, not what is important when viewed in a larger perspective. The practitioners have no taste for application or even for inter-disciplinarity since this takes them away from their own universe. They naturally and honestly try to impose their style and their standards of value upon society, as when they insist on research instead of action, or when they claim urgency for matters whose urgency — that is, importance — is largely self-generated.

For the universities, and for the members of the universities, I have some recommendations, though I put them forward diffidently. In the first place the university must accord the generalist of broad outlook the status and prestige it now confers solely upon the specialist of narrow outlook; and in the second place, the university must rededicate itself to education, including undergraduate education. I realize that the first of these measures is viewed with suspicion by the university. Specialization is "blessed" in the sense that only the specialist knows *what* he is talking about; yet, if only the specialist knows what he is talking about, only the generalist knows *why* he should talk at all.

Can the university combine the point of view of the specialist with that of the generalist? Can it acquire some of the mission-orientation of the large laboratory, yet retain its discipline-orientation intact? Can it truly become interdisciplinary and whole, and thus become congruent with society? Several possibilities for achieving these ends suggest themselves. The university could convert itself into the national laboratory. This is surely going too far, even though mission-oriented institutes are springing up on university campuses, largely I believe in response to the contradictions that I have outlined. The university certainly should not give up the freedom and the individual autonomy of the professor, the freedom and autonomy he cannot enjoy when he enters the mission-oriented institute. Nor should the university give up its discipline-orientation; it is the only institution in our so-

161

ciety in which the disciplines can flourish and can progress along lines set by their own logic. It would be a grave mistake for the university to become the national laboratory.

I would go further. Many of the shortcomings I find in the university are intrinsic characteristics of the university and are hardly susceptible to change. The university loses something unique and precious when it submerges the professor's independence to achieve a common scientific mission conceived by administrators, when it loses its discipline-orientation. But this means simply that some things are not properly done at the university. For example, the "important" problems even in pure science that transcend in difficulty the capacity and style of the university, like studies of genetics involving 200,000 mice, or modern plasma physics, must be done outside the university. Moreover, the basic research that goes to support such activities is properly the business of institutions having such responsibilities. Thus my plea amounts to reasserting the validity of the national laboratory, with its shortcomings that I know so well, as a home for certain kinds of basic and applied research, even as I emphasize the place of the university, with its shortcomings, in the scientific society. The view that federal support of basic research is the university's inalienable right, and that if competition with the mission-oriented institutions arises then the university's is the prior claim (as implied in the Wooldridge report on NIH[21]), ignores the shortcomings of the university in basic research.

There is an appropriate analogy here between the two kinds of institutions: the university and the mission-oriented laboratory. Basic research is supported in the mission-oriented laboratory in order to create an environment sophisticated enough to enable the laboratory to accomplish

[21]"Biomedical Science and Its Administration," A Study of the National Institutes of Health, Report to the President, D. E. Wooldridge, Chairman, U.S. Government Printing Office, Washington, D.C. (February 1965).

162

complicated technical missions. As Harvey Brooks[22] suggested, it ought to be looked upon as a reward for achievement of the laboratory's mission, especially since the basic researcher is thereby given a stake in achievement of the laboratory's mission. Similarly, a case can be made for giving the university, as an institution, support for basic research as a reward for excellence in teaching, since one thereby gives the research professor a stake in the university's mission.

Herein, I believe, lies the salvation of the university because in a sense the university, no less than the laboratory, is already mission-oriented if only it will accept and recognize its traditional mission: education of the young. And just as the mission-orientation of the national laboratory adds point and wholeness to its scientific activity, so pregraduate education ought to give wholeness to the university. Education at the undergraduate level should properly be less professionalized and puristic than it is at the highest levels. Just as, according to older biologists, ontogeny recapitulates phylogeny, so elementary education properly should recapitulate the historic path of a discipline: its connections with other disciplines and with practical purposes, its origin, its scholarship — in short, its place in the scheme of things. If the university takes undergraduate education seriously, and does not look upon it simply as attenuated professional education, the university community will be forced to broaden its outlook. And in the process, as the university re-establishes its relevance to the interdisciplinary real world, it can look forward to an immortality sanctioned not only by its antiquity but also by its dedication to the purposes of the society that now supports it.

[22]"Future Needs for the Support of Basic Research," *Basic Research and National Goals*, p. 108, National Academy of Sciences, U.S. Government Printing Office, Washington, D.C. (1965).

3. *The Federal Grant University and the Federal Laboratory*[23]

I turn now to the interaction between the federal grant university and the federal laboratory. The AEC's National Laboratories, particularly, have always had many points of connection with the universities, and the universities with the national laboratories. But in the past few years, both in this country and abroad, some discord has been injected into this fruitful symbiosis: where cooperation is called for, there is sometimes competition; where trust is needed, there is sometimes mistrust.

What I am referring to is illustrated by my experience during a visit to Pakistan. I was wearing two hats on that trip — one, as chairman of a panel on higher technological education in Pakistan, the other, as an emissary of our Atomic Energy Commission. In the mornings, wearing my education hat, I visited vice-chancellors of Pakistan's universities; in the afternoons, wearing my atomic energy hat, I visited directors of government laboratories. From the vice-chancellors I heard complaints that too much money was going to the government laboratories, money that could be better spent to upgrade research in the universities. From the laboratory directors I heard the reciprocal complaint, that the laboratories stood ready to help the universities by lending

[23]Based on a Commencement Address at The University of Tennessee, Knoxville (March 1965).

164

people to teach or by helping students who wished to do advanced research, but that the universities were too jealous of their prerogatives to allow such cooperation. I have encountered this sort of recrimination in several other parts of the world, both in underdeveloped and in developed countries. And even in our country, such misunderstanding is not unknown. The decision by President Johnson in 1963, rejecting a request by a group of Midwestern universities to sponsor a very large accelerator and urging them instead to cooperate more closely with the Argonne National Laboratory, suggests a lurking rivalry between the university and the federal laboratory.

The source of such rivalry is not hard to determine. When the large government laboratories were organized, mostly during the war, the job of the laboratories hardly intersected the job of the universities; but in the intervening years the distinction between the two has blurred. Both institutions now depend upon the federal government for money to support their research, even to the extent that some universities have become federal grant universities. Both institutions depend upon the same inadequate pool of scientists to conduct their research. Federal money and good scientists are limited, and choices between the federal grant university and the federal laboratory are being made. As usual in such situations, those who are left out — sometimes the university, sometimes the federal laboratory — complain.

Such rivalry, insofar as it exists, is good for neither institution. Instead, we must strengthen existing patterns of cooperation, and we must invent new methods of collaboration that reinforce the essential purpose of each institution. I shall describe some of these newer patterns, and in so doing, give again my view of the purposes of the university and of the federal laboratory.

I have already expressed my belief in the validity — that is, the right to continued public support — of both institutions. The validity of the universities, primarily as centers of education, is attested by their immortality; some of them are almost a thousand years old. We shall always have to preserve, create, and transmit knowledge, and our universities will always be the centers for all of these activities. Of these activities, preserving and transmitting knowledge — that is, education — are, to my mind, the ones that most uniquely characterize the university.[24] Other institutions create knowledge; only the university has a primary responsibility to transmit and preserve knowledge.

By comparison with the universities, the federal laboratories are Johnny-come-latelies; yet in this age of scientific revolution I believe their validity is as solid as that of the universities. The primary purpose of the laboratories is to respond to the large-scale scientific and technical needs of the government, and they will merit continued support so long as they vigorously identify these needs and mobilize around them. Though our nation will always be confronted with technological and scientific problems, some of these problems are hard to identify or to mobilize around, especially those that by virtue of their massiveness or difficulty should be done by the federal laboratories. In this respect, the universities have the advantage: education is a never-ending and rather easily recognized need, and the institutions responsible for education will always merit public support, particularly if they continue to recognize education as a primary function.

This picture of the university and the federal laboratory is schematic, and many details are omitted. Though the primary purpose of the federal laboratory — research on applied problems of national importance, or on basic problems that are too big to be handled in smaller places — differs from the

[24]This view of the proper role of the university is by no means held only by those outside the university. For example, President Kingman Brewster, Jr., of Yale University wrote recently, "The distinctive character of a university is its role in the transmission of learning and communication of the excitement of intellectual, aesthetic, and moral advance." *Ventures,* Magazine of the Yale Graduate School, p. 7 (Spring 1966).

primary purpose of the university, which is education and research in the style of Little Science, many of the secondary purposes of the laboratory and the university overlap. The main focus of the federal laboratory is large-scale, usually applied research, done by teams; yet the federal laboratory also does much small-scale, individual basic research. To carry out their main tasks, for example, the national atomic energy laboratories must investigate many areas of science which bear only indirectly on atomic energy yet which interest many university people. And on the other side, though Little Science is the science of the university, many parts of Big Science, even though they are too expensive to be carried out at the university, also are legitimately of interest to university people. Similarly, though education is primarily the business of the university, it has become an important, though secondary, part of the business of the federal laboratory.

The partial overlap of purpose between university and laboratory has in the past been the basis of fruitful cooperation between these two institutions. For example, in the South it has led to the Oak Ridge Associated Universities, an organization of forty Southern universities that has promoted cooperation between Oak Ridge and the ORAU member universities. ORAU has sponsored many exchanges between university and laboratory. Each summer about fifty faculty members participate in research at ORNL. Many members of the Oak Ridge staff lecture at Southern universities, and some even give full courses. Graduate students come to ORNL to do research. Students from neighboring small colleges who would otherwise not have much contact with research spend some time each summer at the laboratory to acquire firsthand knowledge of how modern research is done. In addition to these ORAU-sponsored ventures, the Oak Ridge National Laboratory has in a less formal way influenced education in the South. Many of its consultants are members of faculties of Southern universities. Each year the laboratory accommodates about fifty cooperative students who spend alternate semesters at the laboratory and at their university. In these ways, both formal and informal, the laboratory has been making a marked and I think a beneficial impact on the universities; I know that the surrounding universities' impact on ORNL has been marked and beneficial.

In the last few years new modes of collaboration have developed which have involved the federal laboratories more directly in education. The possibility of the federal laboratories' participating rather directly in graduate education was discussed in section 11 of the so-called Seaborg Report.[25] The main point of this report is that basic research and graduate science education ought to be closely interwoven. Since the federal government pays for, and therefore has responsibility for, most of the country's basic research, it must bear a responsibility for the institutions in which most of the basic research is done, the universities. This responsibility includes support of administrative overhead and of fellowships and the like: in short, support of science education at the graduate level as well as support of basic research. There are two symmetric routes to achieving excellence in both graduate science education and in basic research. One can start with a good school and build up its basic research, or one can start with a basic research institute and adjoin to it a good graduate school. The former scheme is the one traditionally followed in our country, and is the idea underlying the National Science Foundation's Centers of Excellence program. The latter scheme is illustrated by the conversion of the Rockefeller Institute into the Rockefeller University.

What are the advantages and the disadvantages of making our federal laboratories look very much more like graduate science schools than they now do? The first and primary ad-

[25]"Scientific Progress, the Universities, and the Federal Government," The President's Science Advisory Committee, U.S. Government Printing Office, Washington, D.C. (November 1960). See also Philip C. Ritterbush, "Research Training in Governmental Laboratories in the United States," *Minerva IV*, 186–201 (Winter 1966).

vantage is that we could thereby quickly increase the number of centers of excellence in graduate science. I estimate, for example, that the three large nonsecret atomic energy laboratories — Argonne, Brookhaven, and Oak Ridge — plus the nonsecret parts of Livermore and Los Alamos could accommodate between 1500 and 2000 doctoral candidates in the physical and biological sciences. This represents about 2 per cent of the total graduate student population in the sciences as of 1962.[26]

Moreover, excellent federal laboratories are distributed rather more uniformly through our country than are excellent universities. The Los Alamos Laboratory, in an isolated part of the Southwest, is one of the world's great scientific institutions; many of the strongest space laboratories, such as Marshall in Huntsville or the Manned Space Flight Center in Houston, are in the South. By exploiting the educational possibilities of these institutions, one could rather automatically spread *educational* excellence to regions of the country not presently so endowed. The possibility of thus using the rather accidental location of excellent federal laboratories as a means of achieving better geographic distribution of centers of educational excellence has been recommended by both G. B. Kistiakowsky, President Eisenhower's Science Adviser, and by D. F. Hornig, President Johnson's Science Adviser.

The centers themselves would be improved in many ways. First, the flow of graduate students would add zest and originality to the basic research done at the laboratories. Most scientists never work as hard at any other time in their lives as they do as graduate students; to have graduate students around would raise the accepted norm of hard work at these establishments. At least their presence would sustain a general enthusiasm which is so easy to lose, even in basic re-

[26]"Scientific and Technical Manpower Resources — Summary Information on Employment, Characteristics, Supply, and Training," National Science Foundation NSF 64–28, p. 130, U.S. Government Printing Office, Washington, D.C. (1965).

search, when the research is done year in and year out by the same people, each of whom grows one year older each year. Second, contact with young, beginning students requires the researcher to be a teacher, to order his views, to make sense in detail as well as in general, to scrutinize carefully what he takes for granted. It is easy to be content with a partly thought-out view, to take for granted something one understands only dimly because one doesn't have to go over the point in the detail that is necessary in preparing a lecture. Because of this tendency, chances to make great discoveries are missed.

The main danger in making our large laboratories over into combination basic research institutions and centers of graduate education is that the job of turning out Ph.D.'s might divert the laboratories from their primary purpose of developing nuclear weapons or reactors or rockets. The testing of a nuclear weapon must be done by professionals; one cannot wait until an inexperienced graduate student catches on. In this respect, the two objectives — education and development — tend to be incompatible. Yet even in applied project work graduate students can be surprisingly useful. At Oak Ridge we had for thirteen years a branch of the Massachusetts Institute of Technology Practice School of Engineering. Candidates for M.S. degrees in engineering spent one semester at Oak Ridge doing a succession of small, but useful, development jobs. We found these bright young people very helpful to us, and I think most of them learned a great deal.

But the graduate students would be involved, in any case, mostly with the basic research at the large laboratories, and this work usually has no deadline. Even laboratories with the heaviest program of applications, such as the Atomic Energy Commission's weapons laboratories, do much basic research. In this respect, as I have already suggested, the original distinction between university and federal laboratory — one concerned almost exclusively with basic, and the other with applied, research — is blurred. The basic research sector in the laboratories is so big that the laboratories could offer ample opportunities for thesis work in basic science alone. What the students bring to the basic research in added originality would, I believe, offset any loss in volume of basic research.

The most drastic way to bring the federal laboratories into the mainstream of graduate science education would be to convert them into, say, federal M.I.T.'s. This resembles the pattern that was adopted at Rockefeller; or at the Scripps Institute of Oceanography, which became the nucleus of the University of California at La Jolla. At one time I had advocated such a course, but I now realize that to convert our federal, mission-oriented laboratories into universities would be a mistake. As I argued earlier, the mission-orientation of the federal laboratories is a precious thing. If they became discipline-oriented universities our society would lose what I consider to be an indispensable asset, the scientific institution that aims always to mobilize around the society's compelling sociotechnical problems.

I have therefore come to the view that the laboratories would be much better advised to join with existing educational institutions in setting up joint programs in which both laboratory and university participate, but in which both partners maintain their separate identity and purpose. One plan is to make the very best basic research scientists at the federal laboratory part-time professors at the neighboring university. These individuals, if carefully chosen, could help convert an average institution into a center of excellence. These special professors would spend, say, half their time at the university and half at the laboratory; they would be faculty members who happen to do their research at the laboratory.

Many such schemes have sprung up during the past few years. At my own laboratory, the Ford Foundation has underwritten the initial cost of adding to the faculty at the nearby University of Tennessee some three dozen members of the

Oak Ridge scientific staff. Both laboratory and university were at first concerned as to how the Oak Ridge professors would be accepted by the departments; this, however, proved to be a needless worry. The Oak Ridgers participate in the academic affairs of the university in the same spirit as do the full-time professors. The experiment has been going on for three years now, and is generally judged to be successful.

Encouraged by this success, The University of Tennessee has recently taken still another step toward exploiting nearby Oak Ridge for educational purposes. The university has established a graduate school of Biomedical Sciences next to the Biology Division of the Oak Ridge National Laboratory. The Graduate School is a part of the university, but most of the faculty will consist of Oak Ridge biologists who will become part-time members of the faculty in much the same way as are the Ford Foundation professors. The UT-Oak Ridge Graduate School of Biomedical Sciences resembles other joint institutes such as the UT Space Institute at Tullahoma, next to the Arnold Engineering Development Center, or the University of Alabama Research Institute at Huntsville, next to the Marshall Space Flight Center.

I have touched upon two complementary trends which are influencing both the universities and the laboratories. On the one hand, the universities, in depending more and more on the federal government for the support of their research — in becoming federal grant universities — are coming to resemble ever more closely the federal laboratories. On the other hand, the laboratories, in responding to the intrinsic connection between research and teaching, are coming to resemble the federal grant universities.

This trend toward blurring of our institutions by each taking over historic functions of a related institution is very common today. In industry it is called diversification: a process which turns an oil company first into a chemical company, then into an equipment manufacturer as well, or an airplane company into a submarine company and eventually

172

a chemical company and a personnel broker as well. In universities it is marked by a tendency to do more things that have little to do with education; in laboratories it is marked by a tendency to engage in an ever broadening spectrum of activities that divert the laboratories from their missions.

With respect to the growing resemblance between the federal grant university and the federal laboratory, I have only this to say: the trend is healthy and should be encouraged if it furthers the central purpose of each institution. It is unhealthy and should be discouraged if it compromises the central purpose of either institution. As I have pointed out, the central purposes of these two institutions — education and small-scale research for the universities, and research on national problems and large-scale, interdisciplinary basic science for the federal laboratories — are both valid and necessary. Our society must have both purposes fulfilled not in a way that compromises one purpose by pursuit of the other, but in a way that furthers each as fully as possible.

My own view is that the university's central purpose — education, coupled with Little Science — is compromised if the university becomes indistinguishable from the federal laboratory; and that the federal laboratory's ability to mobilize sharply and decisively around the most urgent national problems declines if the laboratory becomes indistinguishable from the federal grant university. Each must retain its own characteristics; each must maintain its identity and integrity.

Within these limits, fruitful cooperation and reinforcement are possible and desirable. The beauty of the joint Oak Ridge–University of Tennessee arrangement I have described is that each kind of institution — Oak Ridge National Laboratory on the one hand, The University of Tennessee on the other — retains its integrity of purpose, in fact strengthens its ability to fulfill this purpose, and yet reinforces its partner. I would hope that this pattern between these two institutions can serve as a model for other such cooperative ventures between federal laboratories and universities; that such under-

173

takings will spread the tradition of scientific excellence to many educationally deprived areas of the United States; and that both our universities and federal laboratories, far from competing, will continue mutually to reinforce each other to the great advantage of the society that supports them.

INDEX

Abstractness, in mathematics, 44, 45, 151
in science, 44ff
Administrative choices, 87
Ahlfors, L. V., 155n
American Council for Curricular Evaluation, 155
American Electric Power Company, 14
Ammonia, production of with nuclear energy, 9n
Anderson, Clinton P., 134
Anderson, N. G., 108
Applied science, 87ff
as mission overhead, 90
relation to pure science, 148
status in universities, 89, 155
Aqueous Homogeneous Reactor, 129
Argonne National Laboratory, 106, 124, 127, 129, 131, 165, 169
Arnold Engineering Development Center, 172
Artistic truth, 93
Arts, analogy between arts and science, 92
artistic criticism, 120
support of, 94
Associated Midwest Universities, 124n
Atomic Energy Commission, 58, 60, 61, 106, 107, 109, 112, 124, 135, 136

Automatic dialing systems, 4
Automation, 26

Barzun, Jacques, 93n
"Basic Research and National Goals," 98n
Basic science, analogy with arts, 92
as high culture, 36, 91ff
its motivation, 89
in national laboratories, 124, 132
as overhead, 91, 97ff
Battelle Memorial Institute, 126
Behavioral sciences, 77, 81
Bell Laboratories, 92
Bender, M., 33n
Benzer, Seymour, 112
Bethe, Hans, 155, 156n
Bettis Laboratory, 129
Big Biology, 101, 106ff
Biomedical science, 101ff
as applied science, 91, 103
as Big Science, 113
and computers, 109
and engineering, 108
indebtedness to physical sciences, 111ff
institutes for, 110ff
its interdisciplinarity, 108
as Little Science, 107
ripeness of, 104
support of, 101ff
UT-ORNL graduate school of, 172

175

Biosphere, insults to, 32, 107
Birth control, 141
Blanco, R. E., 9, 10
Bohr, N., 49
 Bohr's Aufbau Prinzip, 49
Bonner, James, 6, 13n
Bradbury, Norris E., 129
Brain, as computer, 28
Breeder reactors, 19, 22ff, 130, 137
Brewster, Kingman, Jr., 166n
Bronowski, Jacob, 116, 117, 119
Brookhaven National Laboratory, 123
Brooks, Harvey, 146, 163
Brown, Harrison, 6, 11n, 13n, 21, 23
Brown, K. B., 23
Browns Ferry Nuclear Power Plant, 16
Bureau of the Budget, 68, 69
"Burning the rocks," 22
"Burning the sea," 21
Bush, Vannevar, 60

Calvin, M., 28
Cameron, D. E., 28
Cardinal Steam Plant, 14, 15
Carson, Rachel, 30, 139
CERN, 76
Chain reaction, 128
Citation indexing, 56
Civil defense, 139
Clark, Sir Kenneth, 120
Coal, 12, coal mining machines, 15
 conversion into liquid fuel, 12
 economics of coal-fired power plants, 15
Cockcroft, Sir John, 131n
Cole, Kenneth, 112
Computers, 25
 and biomedical science, 109
 and document retrieval, 56
 as model of brain, 28
 non-numerical uses of, 71

Congressional Appropriations Committees, as instruments of allocation, 68
Conservation of parity, 79
Conway granites, as source of residual thorium and uranium, 23
Cosmology and high-energy physics, 79n
Courant, Richard, 151
Crick, Francis, 112
Criteria for scientific choice, 65ff, 85ff
 external criteria for scientific choice, 72ff, 86
 internal criteria for scientific choice, 71ff, 86
 between science and other activities, 85ff
Curriculum reform, and the universities, 153ff

Darwin, Sir Charles, 7
Darwin, Charles R., and evolution, 50, 51
Davenport, W. H., and "Project Literacy," 54
Defense Documentation Center, 58
"Delegated agent," 61
Department of Defense, 106, 112, 135
Department of Health, Education, and Welfare, 135
Department of the Interior, 133
Desalting the sea, 14, 20, 30, 80, 132, 133, 134, 135
 R. P. Hammond's role, 134
 in Israel, 30
 and redeployment of laboratories, 135
 triagency study, 20
Disciplines, 60
 and universities, 124, 145ff
Document retrieval, 55ff
 centralized depository, 57
 journals, 57
 preprints, 57

Doty, Paul, 112
Dounreay Fast Breeder Reactor, 24
Dresden No. 1 Boiling Water Nuclear Plant, 17

Egyházi, E., 28
Electric automobile, 31, 141
Electrical heating, 13
Electricity, conversion into various products, 8ff
cost of generation, 15, 19ff
importance of cheap, 7ff
Electron microscope, 106
Elementary particles, 48, 148
Elsasser, W. M., 51
Embeddedness of values, 74, 82, 117, 120, 160
Endicott, K. M., 33n
Energy, agency for its development, 137
conversion into various products, 8ff
energy budget for world, 13
fission, 21
fusion, 21, 24, 73, 138
gigantism in, 34
as ultimate raw material, 12
see also Nuclear energy
Entropy, and ethics, 119
and information, 5
Etherington, Harold, 129
Ethical relativism, 121
Ethics, 117ff

Federal grant universities, 124, 164ff
and federal laboratories, 164ff
Feinberg, Gerald, 79n
Fermi, E., 128, 133, 135
Fink, Mary Alexander, 33n
Ford Foundation, 171, 172
Forster, E. M., 26
Fossil fuel reserves, 14
Friedman, D., 31
Fuel cells, 12
Fusion, controlled, 21, 24, 73, 138

Galbraith, John K., remarks on poverty, 36
Gaseous diffusion plants, 18, 20, 108, 128
General Electric, 16, 17, 18, 129
Generalists, 27
as theorists, 48ff
in the university, 161
Government and information, 57ff
Graduate education, and federal laboratories, 166, 168
geographic distribution of, 169
Granitic rocks as source of uranium and thorium, 22, 23, 29

Hahn, O., 133
Hammond, R. P., 8ff, 134
Hanford atomic energy plant, 18, 128
Harwell Atomic Energy Research Establishment, 131n
H-bomb, 33, 129
Heckscher, August, 95
Heisenberg, W., 133
High-energy physics, 60, 75, 77, 78ff, 81
compared with quantum mechanics, 79
and cosmology, 79
as instrument of international cooperation, 80
and other branches of science, 159
in the university, 159
Hoffman, George, and electric automobile, 31
Holmes, J. M., 9
Hornig, D. F., 169
Hoyle, Fred, 40
Hubbert, M. King, 44
Human value, 87, 117ff
and order, 119
and scientific choice, 115ff
Humanists' responsibility, 35
Hyden, H., 28

Hydroelectricity, 13
Hyperbaric x-ray therapy, 103

ICBM, 132
Induction, systematization of, 47ff
Information, 39ff
in biology, 27ff
centers, 51ff
and entropy, 5
explosion, 57, 70
and government, 57ff
and Malthus' second dilemma, 5ff
retrieval, 56
transfer chain, 53
Information center, 51ff
as staff for science, 53
as systematizer of induction, 51ff
Institute for Scientific Information, 56
Institutional choice, 65
Institutional purposes, 144
Interdisciplinary materials research laboratories, 106

Jensen, J. H. D., and shell model of nucleus, 50
Jersey Central Power and Light Company, 16
Joliot-Curie, F., 133
Jordan River Project, 30

Kac, Mark, 41, 151
Keppel, F., 154n
Kerr, Clark, 124
Kistiakowsky, George B., vi, 169
Knolls Atomic Power Laboratory, 129
Korzybski, Alfred, 47
Kuhn, Thomas S., 41n

Laser, 46
Ledley, Robert S., 33n
Leisure and affluent society, 36
Leukemia and viruses, 32, 33, 147

Lewis Laboratory (NASA), 127
Leymaster, Glen R., 72n
Lilienthal, David E., 2
Limits of science, 42
Lincoln Laboratory, 25, 124
Lindsay, R. Bruce, 119, 121
imperative toward order, 119
Liquid fuels, 13, 138
Little Biology, 109
Livermore Laboratory, 129, 169
Los Alamos Scientific Laboratory, 8, 129, 169

McConnell, J. V., 28
McElroy, William D., 157
McNeill, David, 54n
Malthusian dilemmas, 29, 137
first dilemma (energy), 3, 4
second dilemma (entropy), 3, 4, 24, 25, 26, 35, 39, 41
Mars, exploration of, 137
Marschak, Jacob, 10
Marshall Space Flight Center, 126, 172
Mass transportation, 139
Massachusetts Institute of Technology, Practice School of Engineering, 170
Materials Testing Reactor (MTR), 129
Mathematics, 45
abstractness in, 44, 45, 150
and curriculum reform, 154ff
in the university, 154
Mayer, Maria G., 50
Medical education, 72
Medical scintillation spectrometry, 109
Mendeleev as a compactor of literature, 49, 51
Mesic atoms and high-energy physics, 79
Metallurgical Laboratory (wartime plutonium laboratory) 126, 127, 129, 131
Missions, 60
and government agencies, 135
and national laboratories, 123ff

178

possible future missions, 137
and society, 145
and universities, 145
Mitchell, A., 95n
Molecular biology, assessment of,
    77
Molten fluorides in reactors, 147

National Academy of Sciences,
    156
National Advisory Committee for
    Aeronautics, 126
National Aeronautics and Space
    Administration, 112
National Bureau of Standards,
    126
National Cancer Institute, 109
National Humanities Foundation,
    95
National Institutes of Health,
    111, 123, 144, 162
National laboratories, v, 123ff,
    126ff, 164ff
and basic research, 124, 131,
    170
and education, 167ff
mission of, 123ff
purpose of, 166
redeployment of, 136
and society, 142
successes of, 127
and technological risks, 130
trend toward disciplinarity in,
    132
and universities, 164ff, 171
National prestige as a criterion of
    merit, 76
National Science Foundation, 60,
    95, 98, 99
center of excellence program,
    168
relation to other agencies, 98
National Standards Reference
    Data System, 52
Nautilus, development of, 129
Nirenberg, M., and genetic code,
    78
Northeast power failure, 34

Nuclear chemistry, 46
Nuclear energy, 16ff, 77, 80, 132,
    133, 138
and air pollution, 141
raw materials for, 21ff
Szilard's role in, 133
Nuclear energy revolution, 14ff,
    21ff
Nuclear-powered airplane, 129
Nuclear reactors, 17ff
containment, 32
development at national labo-
    ratories, 128ff
disadvantage of size, 33
energy cost, 20
fuel cycle, 18, 19
safety, 34
scaling laws, 17, 18
see also Breeder reactors
Nuclear weapons, as deterrent, 29
development at national labo-
    ratories, 128ff

Oak Ridge Associated Univer-
    sities, 167
Oak Ridge Institute of Nuclear
    Studies, 53
Oak Ridge National Laboratory,
    v, 1, 8, 9, 31, 53, 106, 107,
    108, 110, 123, 127, 128,
    129, 131, 140, 143, 147,
    167, 169, 170, 172, 173
Ochoa, S., and genetic code, 78
Office of Civil Defense, 139
Office of Saline Water, 133
Office of Science and Technology,
    68
Oppenheimer, J. Robert, 128, 159
Order, and values, 119
Organ transplantation, 103
Oyster Creek Boiling Water Re-
    actor, 16, 17, 19

Pakistan, Indus Valley Basin, 64,
    82
rivalry between universities
    and laboratories in, 164
Parkinson, C. Northcote, 142

Particles, strange, 78
Pasternack, Simon, 62n
Pauli, W., 46
Peierls, R., 133
Permuted-title indexing, 56
Petrochemicals, 13
Piore, E. R., v
Polanyi, Michael, 93, 118, 119
Polio vaccine, 76
Political choices, 87
Pollution, 30, 139
Population, 2ff
    control of, 87, 139, 141, 156
    see also Malthusian dilemmas
President's Commission on Heart
    Disease, Cancer and
    Stroke, 104
President's Science Advisory
    Committee, vi, 52, 61, 68,
    69, 134, 139, 143, 144
Pressurized water reactors, 129
Price, Derek de Solla, 102
"Project Literacy," 54
Putnam, Palmer, 6

Race relations, 156
Radiation Laboratory, 128
Radioactive wastes, 31
Radioisotopes, 129
Rapoport, A., 117, 118, 119
Rauscher, Frank J., 33n
Redeployment, of laboratories
    and agencies, 132ff
Relatedness, and embeddedness
    of values, 116
Religion, analogy with science, 95
Republic of the arts, 94
Republic of science, 93, 94, 118
Research institutes, 110, 123ff
    for biomedical research, 110
    interdisciplinarity in, 110
Rickover, H. G., 129, 135
Ritterbush, Philip C., 168n
Rockefeller University, 168, 171
Rossi, Peter H., 110
Ruddle, Frank H., 33n
Ruhe, C. H. William, 72n

Russell, W. L., and mouse genetics, 107

Salmon, R., 9
SAM laboratories, 128
Savannah River Plant, 18
Schurr, S. H., 10
Science, abstractness of, 44ff
    Big, 39ff
    advantage of bigness in, 77
    as codifier, 152
    extrinsically motivated, 147
    fan clubs, 96
    international cooperation in,
        76
    intrinsically motivated, 147
    limits of, 42ff
    Little, 39, 83, 107ff
    its motivation, 89
    public support of, 72
    purity of, 147, 148
    as search, 152
    specialization of, 42ff
    teaching of, 46ff, 63
    unity of, 75, 116
    its validity, 59, 70
    see also Applied science, Basic
        science, Biomedical science
"Science, Government, and Information," 42n, 61
Scientific choice, 65ff, 85ff
    and human values, 115ff
Scientific choice, criteria for, 65ff
    external criteria for, 72
    internal criteria for, 71
    and scientific merit, 73ff
    and social merit, 75ff
    and technological merit, 72ff
Scientific communication, 39ff
Scientific community, its hierarchical structure, 47, 48
Scientific creation, 43
Scientific criticism, 68ff, 93, 115
Scientific fashions, 45ff
Scientific information "crisis," 26
Scientific literature, 39ff, 70, 118
    compaction of, 49ff

importance of "refereed," 59, 63
and validity of science, 70
Scientific merit, 75, 116
Scientific parsimony, 120
Scientific policy, 67
Scientific taste, 66, 116
Scientific-technical revolutions, 1ff
Scientific truth, 93
Scientific value, 115
Scientists' responsibility, for communication, 63ff
social, 35, 155
Scripps Institute of Oceanography, 171
Seaborg Report, 168
Semenov, N. N., 95
Shaw, George Bernard, 95n
Shell model of nucleus, 50ff
Shippingport Nuclear Power Plant, 129
Silver, L. T., 11n
SKETCHPAD, 25
Sloan-Kettering Institute, 111
Smith, McGregor, 32n
Social problems and redeployment of laboratories, 140
Solar energy, 13
Solyom, L., 28
Space, exploration of, 77, 81
Space heating, 12
Space Technology Laboratories, 123
Specialists, 26
in the university, 161
Spin-off, 73
Sporn, Philip, 15n
Sulfur dioxide as a pollutant, 30
Szilard, L., 5, 133, 135

Taube, Mortimer, 70
τ-particle paradox, 79
Taylor, A. N., 72n
Technical writing, 53ff
"Technological fix," 141, 158n
Teller, Edward, 89, 129, 155

Tennessee Valley Authority (TVA), 10, 12, 16, 35
Browns Ferry Reactor, 17
Kingston Steam Plant, 30
Thermodynamic revolutions, 2ff
Thermodynamics, 148
first law, 5
second law, 5, 149, 152
Thermonuclear plasma, 150
Thirring, Hans, 6
Tillich, Paul, 36
Tipner, Anne, 72n
Toulmin, Stephen, 87, 88n

Ulam, Stanislaw, 129
Ullmann, J. W., 9
Union Carbide Corporation, 128
Unitrains, 15
Universities, 124ff, 145ff, 164ff
and academic freedom, 110, 161
and applied science, 89, 155
and big science, 167
conflict between research and teaching, 153
and curriculum reform, 153
and disciplinarity, 110, 124, 145ff
educational mission of, 161
and generalists, 161
influence on society, 159ff
and little science, 167
purpose of, 166
rationality of, 156
social structure of, 150
trend toward purity in, 147ff
The University of Tennessee, 171, 172
Upton, A. H., 32n
Urban renewal, 139

Vick, F. A., 131n
von Neumann, John, 33, 34, 46, 74, 149, 150, 151

Water-for-Peace, 9
Watson-Crick model, 111
Weather modification, 33

Weinberg, A. M., 1n, 6n, 55n,
65n, 79n, 85n, 98n, 101n,
110n, 113n, 115n, 126n,
145n, 164n
Weir, John, 6, 13n
Weisskopf, V. F., 79n
Wells, H. G., 2, 7
Westinghouse Corporation, 17,
18, 129
Wiener, Norbert, 25
Wiggins, Walter S., 72n
Wigner, Eugene P., 42, 44, 47,
119, 133, 139

Wooldridge Report, 162
World Health Organization, 111
Wynne-Edwards, V. C., 4n

Xenon compounds, 132

Y-12 Electromagnetic Separations
Plant, 128
Yankee Atomic Electric plant, 16
Yuan, Luke C. L., 159n

Zinn, W. H., 129
Zonal centrifugation, 108, 132